STUDY GUIDE FOR

Berry and Lindgren's

STATISTICS

THEORY AND METHODS
SECOND EDITION

Bernard W. Lindgren
University of Minnesota

Donald A. Berry
Duke University

Duxbury Press
An Imprint of Wadsworth Publishing Company
I(T)P® **An International Thomson Publishing Company**

Belmont • Albany • Bonn • Boston • Cincinnati • Detroit • London • Madrid • Melbourne
Mexico City • New York • Paris • San Francisco • Singapore
Tokyo • Toronto • Washington

Table of Contents

Preface

The Study Guide for *Statistics: Theory and Methods, Second Edition* consists of two parts: additional problems with solutions (Part I) and solutions to the "Review Problems" that appear at the end of each chapter of the text (Part II). Each part is designed to meet a need that we've heard expressed by students.

A common complaint about textbooks (ours and some others) is that there are not enough examples worked out in detail. This is why we give, in Part I, additional worked-out examples (problems with solutions) for practice. After going over the examples in the text proper, students might first try to solve these additional problems on their own, turning to our solutions when stuck, or to check against their own solutions. Next come the starred problems in the text, those for which answers (but not solutions) are included in the text. We have left some problems without answers (those not starred) for instructors who wish to assign such as homework.

Invariably, when an exam is imminent, students have asked us if they can have a copy of an old exam, as a means of review and as a way of seeing (so they think) what kind of problems we ask on an exam. The review problems at the end of each chapter originated as problems from old exams, although in some cases they have been extended somewhat, the student being free of the time constraint of an exam. These are different from the problems given at the ends of sections, in that they are not presented as being associated with a particular section. So we think of the review problems at the chapter ends as just that—problems that provide a way of reviewing the material covered, reminding them of the various ideas and techniques without telling them just which ones are appropriate. For study in preparation for an exam, it may not be sufficient to have just the answer to a problem; hence the solutions we give as Part II of this guide.

D. A. B
B. W. L.

Part I

Additional Problems
with Solutions

CHAPTER 1: Additional Problems

Sections 1.1-1.2

1-A　Give an appropriate sample space for these experiments:
(a) Tossing a penny and a die together.
(b) Selecting a family at random and asking whether they are watching TV and, if so, which channel.
(a) Spinning a pointer that has a continuous scale showing the angle (measured from a reference position) at which it stops.

1-B　In each of the following cases, reproduce the Venn diagram of Figure 1. In that figure, let E denote the set of points in the circle and F, the set of points in the triangle, and shade the given event:

(a) $E^c F$　　　(b) $E \cup F^c$　　　(c) $E \cup E^c F$　　　(d) $EF \cup E^c F^c$.

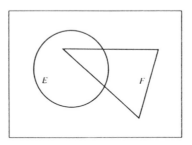

Figure 1

1-C　Consider the sample space $\Omega = \{a, b, c\}$.
(a) List all possible events defined on this sample space.
(b) Define events $E = \{a, b\}$, $F = \{a, c\}$, $G = \{b, c\}$ and list the outcomes in each of the following:
　　(i) EFG^c　　　(ii) EFG　　　(iii) $E \cup FG$　　　(iv) $EF^c \cup EF$.

1-D　Simplify each of the following:
(a) $(E \cup F)^c EF$.
(b) $(A^c \cup B^c A)^c$.

1-E Give an argument, not depending on a Venn diagram, that would establish the distributive law: $A(B \cup C) = AB \cup AC$.

1-F In a draft lottery, each of the 366 dates in a year is marked on a capsule, and a capsule is selected at random from the 366. Find the probability
(a) that it is a December date.
(b) that it is not a date in April or May.

Section 1.3

1-G In the roll of a pair of ordinary dice, the outcomes are ij, where i is the outcome of one die, and j is the outcome of the other. (We keep track of which is which.) The complete list of ij's was shown in Example 1.4a, and we assumed there that the 36 pairs are equally likely. Find the probability that
(a) the sum $i + j$ is 6.
(b) the second die shows 5.
(c) i and j differ by at least 3.
(d) the larger of i and j is 4.

1-H Suppose you hear that the odds against a horse's winning a race are 8 to 3. What is the implied probability of its winning?

1-I Given a population of N equally likely outcomes, show by counting outcomes that for any event E, $P(E^c) = 1 - P(E)$.

Section 1.4

1-J In how many ways can ten distinguishable coins fall?

1-K A menu has 3 appetizers, 2 soups, 4 salads, 6 entrees, and 5 desserts. How many different meals (consisting of appetizer, soup, salad, entree, and dessert) are possible?

1-L Count the possible permutations of the letters in the word
(a) *banana.*   (b) *statistics.*

1-M How many connecting cables are needed to link each pair of nine offices directly?

1-N I want to compare the effectiveness of three treatments, and have 12 lab animals that can be used in a test. A standard method is to divide the 12 randomly into 3 groups of 4, applying treatment A to one group, B to another, and C to the remaining group. In how many distinct ways can animals be assigned to treatment groups?

Section 1.5

1-O In the draft lottery of Problem 1-F, with selections made one at a time and at random, find the probability that the first 3 selections include exactly 2 December dates.

1-P A committee of 4 is to be selected at random from a group consisting of 10 labor and 5 management representatives. Find the probability that the committee selected includes
(a) 2 representatives from labor and 2 from government.
(b) at least 1 representative from each group.
(c) the chair of each delegation, labor and management.

1-Q In a sequence of four random digits, what is the probability that the digits are all distinct?

1-R Find the probability that a five-card poker hand has "one pair," a term which means that the other three cards must be of three different denominations—and different from the denomination of the pair.

1-S A class of 15 students is to be divided into 3 groups of 5, each group to study and discuss the same topic.
(a) How many different ways can this be done?
(b) What is the probability that a particular student and his girlfriend end up in the same group?

1-T A lot (population) of 12 articles includes 9 that are good, 2 that have only minor defects, and 1 that has a major defect. We sample three from the lot, at random, without replacement.
(a) Give the number of possible (unordered) samples.
(b) Find the probability that all 3 are good.
(c) Find the probability that the sample includes 1 of each kind (1 good, 1 with only minor defects, 1 with a major defect).

1-U Three couples are to be seated at random at a round table. Find the probability that Mr. and Mrs. A are not seated next to each other.

Section 1.6-1.7

1-V When you expand the multinomial $(x + y + z)^9$, you'll find several terms of the form $x^3 y^4 z^2$. How many?

1-W Write out the expansion of $(x + y + z)^4$.

1-X Show the following identity:

$$\binom{n+1}{a,\ b,\ c} = \binom{n}{a-1,\ b,\ c} + \binom{n}{a,\ b-1,\ c} + \binom{n}{a,\ b,\ c-1},$$

where $a + b + c = n + 1$.

1-Y Consider this experiment of chance: Toss a thumb tack in the air and see whether it falls with point straight up (U) or with point down (D). There is no reason to expect that these outcomes are equally likely. Let p denote the probability $P(U)$, and let $q = 1 - p = P(D)$. Now suppose three such tacks are tossed. One way of defining a sample space is in terms of the number of tacks that are point up: $\Omega = \{0, 1, 2, 3\}$. Suppose we assign these probabilities to the outcomes: q^3, $3q^2p$, $3qp^2$, p^3, respectively.
(a) Show that these probabilities sum to 1.
(b) find the probability (in terms of p) that at least two of the three tacks land with point up.

1-Z Property (ix) of §1.7: $P(E \cup F) = P(E) + P(F) - P(EF)$, was left as an exercise for the reader. Show that it follows from the axioms for probability [(i)-(iii)].

1-AA Given $AB = \emptyset$, $P(A) = .3$, $P(B) = .5$, find $P(A^c B^c)$.

Chapter 1: Solutions

1-A **Solution:**
(a) The most detailed sample space consists of all pairs of an outcome for the penny and an outcome for the die:

$$\Omega = \{\text{H1, H2, H3, H4, H5, H6, T1, T2, T3, T4, T5, T6}\}.$$

(b) This depends on the city, and whether the viewer has cable TV. In Minneapolis, ignoring cable:

$$\Omega = \{\text{None, 2, 4, 5, 9, 11, 17, 23, 29, 41}\}.$$

(c) If we measure the angle θ in degrees, $\Omega = [0, 360)$, the interval $0 \le \theta < 360$; if in radians, $[0, 2\pi)$. If we round to the nearest degree, then $\Omega = \{0, 1, 2, \dots, 359\}$.

1-B **Solution:** See Figure 2 (next page).
(a) $E^c F$ consists of the points that are in F but not in E.

(b) $E \cup F^c$ consists of all points except those not in E and those in F—that is, those in $E^c F$. This is the complement of the set in (i).

(c) $E \cup E^c F$ consists of points that are either in E or in its complement as well as in F; the events E and $E^c F$ *partition* the union $E \cup F$.
(d) The points in $E^c F^c$ are not in E and not in F; they are outside both the circle and the triangle. The union we want adjoins that region to the intersection of E and F.

1-C **Solution:**
(a) For each of a, b, c, we decide whether or not to include that point, so there are $2^3 = 8$ ways to form a subset. The subsets are $\emptyset$, $\{a\}$, $\{b\}$, $\{c\}$. $\{a, b\}$, $\{a, c\}$, $\{b, c\}$, and $\{a, b, c\} = \Omega$.

(b) (i) $EF = \{a\}$, and $G^c = \{a\}$, so their intersection is also $\{a\}$.

(ii) None of the sample points is in all three events, so $EFG = \emptyset$.

(iii) The intersection of F and G is $\{c\}$; adjoining $\{a, b\}$ yields Ω.

(iv) The distributive law says $EF^c \cup EF = E(F^c \cup F) = E\Omega = E$. Or, $EF^c = \{b\}$, and $EF = \{a\}$, so their union is $\{a, b\} = E$.

1-D **Solution:**
(a) DeMorgan's law says that $(E \cup F)^c = E^c F^c$, so the given expression is $E^c F^c EF = E^c F^c FE = E^c \emptyset E = \emptyset$.

(b) Again apply DeMorgan's law, and then the distributive law:

$$(A^c \cup B^c A)^c = (A^c)^c (B^c A)^c = A(B \cup A^c) = AB \cup AA^c = AB \cup \emptyset = AB.$$

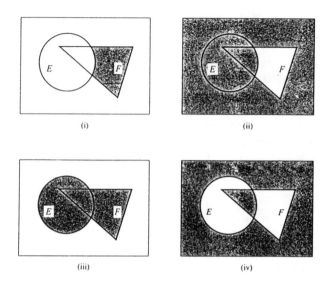

(i)   (ii)

(iii)   (iv)

Figure 2

1-E   Solution:
(1) Suppose an outcome ω is in $A(B \cup C)$; this means that it is in both A and the union $B \cup C$. So, it is in either A and B, or in A and C—and this describes the r.h.s., $AB \cup AC$. (2) Suppose ω is in $AB \cup AC$; this means that it is either in both A and B, or in both A and C. Either way, it is in A. If it is in AB it is in B, and if it is in AC it is in C, so it is in the union of B and C, hence also in A and that union—the l.h.s. Since every point in the l.h.s. is in the r.h.s. and conversely, the l.h.s. and the r.h.s. are the same event—they are "equal."

To show the distributive law for an arbitrary finite union:

$$A(B_1 \cup \cdots \cup B_n) = AB_1 \cup \cdots \cup AB_n,$$

we'd use induction: The crucial step is to write

$$B_1 \cup \cdots \cup B_{n+1} = (B_1 \cup \cdots \cup B_n) \cup B_{n+1}$$

and apply the law for the case of two events to the events on the right:

$$A(B_1 \cup \cdots \cup B_{n+1}) = A[(B_1 \cup \cdots \cup B_n) \cup B_{n+1}]$$
$$= A(B_1 \cup \cdots \cup B_n) \cup AB_{n+1}.$$

Now, if the law holds for n, the A of the first "term" on the right can be applied to each B in parentheses:

$$A(B_1 \cup \cdots \cup B_{n+1}) = [AB_1 \cup \cdots \cup AB_n] \cup AB_{n+1},$$

which says that the law holds for $n+1$. We know it holds for $n = 2$, and hence, for $n = 3$; then, also for $n = 4$, etc. to any finite n.

1-F   Solution:
(a) The 366 dates are assumed to be equally likely. There are 31 days in December, so the probability is 31/366.

(b) There are $366 - 61$ dates in April and May: 305/361.

1-G   Solution: (See Figure 3)
(a) There are 5 pairs with sum 6: (51, 42, 33, 24, 15), marked as (i) in the figure, so the answer is 5/36.

(b) There are 6 pairs in which the second die is 5, marked (ii) in the figure. The answer is 6/36. [This agrees with what you might have thought the answer should be; but the statement of the problem defines the probability model in the sample space of pairs, and in this model, $P(\text{2nd} = 5) = 6/36$ is a consequence.]

(c) There are 12 pairs with this property, marked (iii) in the figure. The desired probability is 12/36 or 1/3.

(d) There are 7 pairs in which the larger number is 4, marked with (iv) in the figure: 7/36 is the answer.

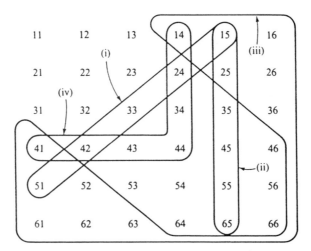

Figure 3

1-H   Solution:
The odds in favor of the horse's winning are 3 to 8; to convert these odds to probability, think of a lottery with 3 chances corresponding to a win and 8 chances corresponding to a loss—11 chances in all. The probability of a win is then 3/11 (3 chances in 11). [Caution: The odds you will hear quoted are "betting odds," giving the ratio of what you put up to what the bookie will put up. This ratio may or may not correspond to what is implied by your (or his) personal probability of a win.]

1-I   Solution:
Suppose the number of outcomes in E is k. The number in E^c must be $N - k$. So,
$$P(E^c) = \frac{\#(E^c)}{\#(\Omega)} = \frac{N-k}{N} = 1 - \frac{k}{N} = 1 - P(E).$$

1-J   Solution:
Each coin can fall in two ways. According to the multiplication principle of §1.4, the number of outcomes (in a list that distinguishes the coins) is $2 \times 2 \times \cdots \times 2 = 2^{10} = 1024$. However, if one were only interested in how many of the coins show heads, the list $\{0, 1, 2, ..., 10\}$ would be adequate as a sample space.

1-K   Solution:
Applying the multiplication principle, we find the number of possibilities as the product of the numbers of choices in the various categories:
$3 \times 2 \times 4 \times 6 \times 5 = 720$.

1-L   Solution:
(a) There are 3 a's, 2 n's, and a b. If they were all distinct, the 6 letters could be arranged in 6! distinct ways. But because there are multiplicities of indistinguishable letters, this must be divided by 3! and by 2!, to yield $6!/12 = 60$, as the proper count. [That is, the letters in each of the 60 ways can be rearranged in $3! \times 2! = 12$ ways, which would be different if the letters were distinguished (say, by subscripts); so the count for all ways when the letters are distinguishable is $60 \times 12 = 720$.]

(b) This time there are 3 s's, 3 t's, 2 i's, 1 c and 1 a—10 in all. The distinct arrangements number $\frac{10!}{3!3!2!} = 50{,}400$.

1-M   Solution:
One cable is needed for each pair of offices; there are as many pairs of offices as there are ways of choosing 2 from 9 things: $\binom{9}{2} = 36$.

1-N **Solution:**
Do the assignment in steps: First choose 4 from the 12 animals to assign to A, then 4 of the remaining 8 to B; the remaining 4 automatically get treatment C: $\binom{12}{4} \cdot \binom{8}{4} = 34650$.

The division into three groups is also counted as $\binom{12}{4,\ 4,\ 4} = \frac{12!}{4!4!4!}$, which yields the same answer.

1-O **Solution:**
The number of distinct combinations of 3 is $\binom{366}{3}$; of these, the number in which 2 are chosen from the 31 December dates and the remaining 1 from the remaining 335 non-December dates is $\binom{31}{2} \times 335$. The probability we want is the ratio:

$$\frac{\binom{31}{2} \times 335}{\binom{366}{3}} = \frac{335 \cdot 31 \cdot 30/2!}{366 \cdot 365 \cdot 364/3!} \doteq .019$$

1-P **Solution:**
There are $\binom{15}{4}$ or 1365 possible selections of 4, equally likely (this is what we mean by "selected at random"). For any given type of committee, we count the number of that type and divide by 1365:

(a) To form such a committee, we choose 2 from the 10 labor *and* 2 from the 5 management representatives. We can do both of these things in $\binom{10}{2}\binom{5}{2} = 45 \times 10 = 450$ ways. The probability is $450/1365 = 30/91$.

(b) The complement of the given condition is that the committee consists of either all labor *or* all management people. The number with all labor or all management is $\binom{10}{4} + \binom{5}{4} = 210 + 5$. (Note: "or" becomes +.) So

$$P(\text{at least one of each}) = 1 - P(\text{all labor or all management})$$

$$= 1 - \frac{215}{1365} \doteq .842.$$

(c) If it is given that the chairs are are to be on the committee, we have only 2 more members to select, and these come from the other 13 people:

$$P(\text{chairs included}) = \frac{13 \cdot 12/2}{1365} \doteq .057.$$

1-Q **Solution:**
To construct a sequence with distinct digits, we choose the first as one in 10, the second as one in 9 (not yet chosen), the third as one in 8 (not yet chosen), and the fourth as one in 7: $(10)_4 = 10 \cdot 9 \cdot 8 \cdot 7$ ways. Without the restriction "distinct," there are $10 \cdot 10 \cdot 10 \cdot 10$ ways. So,

$$P(\text{all distinct}) = \frac{10 \cdot 9 \cdot 8 \cdot 7 \cdot}{10 \cdot 10 \cdot 10 \cdot 10} = .504.$$

1-R   Solution:

As in Example 1.4f, there are 2,598,960 distinct five-card hands. These are equally likely, an assumption implied by the phrase "random deal." To count the number of hands with just one pair, we form such a hand in steps: (1) Pick the denomination of the pair—13 ways; (2) pick 2 of the 4 cards of that denomination—$\binom{4}{2}$ ways; (3) pick three other denominations from the other 12—in one of $\binom{12}{3}$ ways; (4) select one card from each of the 3 sets of 4 cards of those denominations—$4 \times 4 \times 4$ ways. Multiply:

$$13 \times \binom{4}{2} \times \binom{12}{3} \times 4^3 = 1,098,240.$$

The ratio of this to the number of possible hands is about .423.

There are many ways of getting the wrong answer; a student favorite is

$$13 \times \binom{4}{2} \times 48 \times 44 \times 40 = 6,589,440.$$

They reason that after picking the pair, there are 48 cards left for the 3rd card, then 44 for the 4th card, and 40 for the 5th card. However, this would count J-10-4, for instance, as a different set of 3 cards than 4-10-J. So the count is 6 times too large (counting J-10-4 as 3! or six possibilities instead of the correct 1). Dividing it by 6 yields 1,098,240.

1-S   Solution:

(a) The number of ways of dividing 15 into 3 distinct groups is $\binom{15}{5,\,5,\,5}$ or $15!/5!^3 = 756,756$. But, according to the statement of the problem, the groups are *not* distinguished—they have the same charge. So we must divide the count by 3!, obtaining 126,126.

(b) If we take the sample space as the number of divisions counted in (a), the numerator is a count of the number of those divisions in which the two students end up in the same group. We'll count, forming a group by first putting the two students in a group. Whichever it is, we need three more students for that group, selected in $\binom{13}{3} = 286$ ways; then, we need to subdivide the remaining 10 into two undistinguished groups: $\binom{10}{5} \div 2$ or 126 ways. The count is $286 \times 126 = 36036$. The probability we want is $36,036/126,126 = 2/7$.

Having seen the simple-looking answer, we'd surely wonder if there is an easier way. There is: Wherever the young man is assigned, what are the chances of his being joined by his friend? There a 14 other seats, 4 of which are in the same group—4 chances in 14, or probability 2/7.

1-T   Solution:

(a) This is just the number of combinations of 3 from 12: $\binom{12}{3} = 220$.

(b) The number of combinations in which all are from the 9 good ones is $\binom{9}{3} = 84$, so the probability asked for is $84/220 = 21/55$.

(c) For the numerator, we choose 1 from 9, 1 from 2, and one from 1: $9 \times 2 \times 1 = 18$ ways. The probability is 18/220.

1-U Solution:
Mr. A must sit somewhere—seat him first. Then, if seating is random, the probability that Mrs. A sits next to him is 2 (the number of places next to him) divided by 5 (the number of empty places).

1-V Solution:
The number of ways of picking 3 factors from which to use the x, times the number of ways of picking 4 factors (from the other 9) from which to use the y:

$$\binom{9}{3}\binom{6}{4} = \binom{9}{3,\ 4,\ 2} = 1260.$$

1-W Solution:
There are $3^4 = 81$ terms, before collecting "like" terms. The number of the type x^3z, for example, is calculated as in 1-V: $\frac{4!}{3!1!}$; the number of the type x^2z^2 is $\frac{4!}{2!2!}$; and so on:

Type of term	Number of that type	Total
x^4, y^4, z^4	1 each	3
$xz^3, x^3z, y^3z, yz^3, xy^3, x^3y$	$4!/3! = 4$ each	24
x^2yz, xy^2z, xyz^2	$4!/2! = 12$ each	36
x^2y^2, x^2z^2, y^2z^2	$4!/(2!2!) = 6$ each	18

Total: 81

So the expansion is

$$x^4 + y^4 + z^4 + 4(xz^3 + x^3z + yz^3 + y^3z + xy^3 + x^3y)$$
$$+ 12(x^2yz + xy^2z + xyz^2) + 6(x^2y^2 + x^2z^2 + y^2z^2).$$

1-X Solution:
We write each term on the right in terms of factorials and factor out common factors:

$$\frac{n!}{(a-1)!b!c!} + \frac{n!}{a!(b-1)!c!} + \frac{n!}{a!b!(c-1)!}$$

$$= \frac{n!}{(a-1)!(b-1)!(c-1)!}\left\{\frac{1}{bc} + \frac{1}{ac} + \frac{1}{ab}\right\}.$$

The quantity in brackets can be written with a common denominator:

$$\left\{\frac{1}{bc} + \frac{1}{ac} + \frac{1}{ab}\right\} = \frac{a+b+c}{abc}.$$

So the right-hand side becomes

$$\frac{n!(a+b+c)}{a!b!c!} = \frac{(n+1)!}{a!b!c!} = \binom{n+1}{a,\ b,\ c}.$$

Note: This identity can be interpreted as follows. To take a special case, suppose we want to arrange 4 A's, 3 B's, and 2 C's in a sequence; the number of distinct arrangements is $\binom{9}{4,\ 3,\ 2}$. The arrangement can also be carried out in this sequence of steps: Either start the arrangement with an A, in which case there are $\binom{8}{3,\ 3,\ 2}$ ways to finish it; or start it with a B, in which case there are $\binom{8}{4,\ 2,\ 2}$ ways to finish; or start it with a C, in which case there are $\binom{8}{4,\ 3,\ 1}$ ways to finish. (This verifies the identity when $a = 4$, $b = 3$, and $c = 2$.)

1-Y   Solution:
(a) The sum of the assigned probabilities is $q^3 + 3q^2p + 3qp^2 + p^3$, which is what you get in expanding the binomial $(p+q)^3$. Since $p + q = 1$, the sum of the probabilities is 1.

(b) "At least 2" means (in this case) either 2 or 3; so we add ("or") the probabilities of 2 and of 3: $3qp^2 + p^3 = 3p^2 - 2p^3$.

1-Z   Solution:
Observe first that $E(E^cF) = \emptyset$, so that $E \cup E^cF$ is the union of disjoint events. Now apply the distributive law (set union over set intersection):

$$E \cup E^cF = (E \cup E^c)(E \cup F) = \Omega(E \cup F) = E \cup F.$$

Then, by the additivity axiom (iii), we have

$$P(E \cup F) = P(E) + P(E^cF) = P(E) + [P(F) - P(EF)],$$

since $P(F) = P(EF) + P(E^cF)$ (law of total probability).

1-AA Solution:
Since $A^cB^c = (A \cup B)^c$, $P(A^cB^c) = 1 - P(A \cup B) = 1 - P(A) - P(B) = .2$.
$[P(AB) = 0$ because it is given that A and B are disjoint.]

CHAPTER 2: Additional Problems

Sections 2.1-2.2

2-A In Example 1.7b, we gave probabilities for the number of tosses of a coin required to obtain heads for the first time. Let X denote this number of tosses—a random variable. Its p.f., as given by the probabilities in the example, is $f(k) = 1/2^k$, $k = 1$, 2, ...
(a) Find $P(X > 2)$.
(b) Find $P(X$ is odd$)$.

2-B Find the probability function of Y, the number of hearts in a random selection of two cards from a standard deck of playing cards.

2-C Consider the sample space of outcomes $w = (i, j)$, where i and j are digits, that is, integers from 0 to 9, and define $P[(i, j)] = .01$ for each possible outcome. Now define $Y(w) = i + j$, and $Z(w) = i - j$, and find the p.f. of each variable.

2-D In Problem 1-36 we selected three socks at random from a drawer that contained 10 socks: 2 red, 4 green, 4 blue. The sample space is the set of $\binom{10}{3} = 120$ possible selections. On this sample space, define random variables R, the number of red socks and G, the number of green socks among the 3 selected.
(a) Construct a table showing $f_{R,G}(r, g)$, the joint p.f. of R and G.
(b) Give the marginal distributions of R and G.
(c) Find $P(R + G = 2)$.
(d) Find $P(R = G)$.

Section 2.3

2-E A digit is selected at random from 0, 1, 2, ..., 9. Consider the following events: $E = [$the digit is even$]$, $T = [$the digit is a multiple of 3$]$, $F = [$the digit is a multiple of 4$]$. Find the following probabilities:
(a) $P(E \mid F)$ (c) $P(E \mid T)$ (e) $P(T \mid F)$
(b) $P(F \mid E)$ (d) $P(T \mid E)$ (f) $P(F \mid T)$

2-F Life tables used by the National Center for Health Statistics showed that the proportion of black males born in 1973 who would live to age 20 is .947. The proportion who live to age 65 was given as .499. On the basis of these proportions, find the probability that a black male who was 20 years old in 1993 would reach age 65.

2-G   Suicides in the U.S. in a particular year were classified according to the means used, for males and for females, the results given as percentages:

Category	Male	Female
Explosives, firearms	63.5	35.7
Poison	15.8	43.0
Hanging, strangulation	14.3	11.5
Other	6.3	9.75
Number:	18,595	7,088

Find the proportion of suicides that were
(a) male.
(b) by poison.
(c) female, if by poison.
(d) male, if by explosives or firearms.

2-H   A carton of 10 items includes 2 that are defective (D) and 8 that are good. We select 2 at random, one at a time and without replacement. Find the probability that
(a) the first is defective and the second is good (DG).
(b) the first is good and the second is defective (GD).
(c) one is defective and the other is good (DG or GD).
(d) both are defective (DD).
(e) neither is defective (GG).

2-I   Two cards are picked at random from a standard deck. Find the probability that
(a) both are hearts, given that both are red.
(b) both are face cards, given that they are of the same suit.
(c) both are red, given that the second one is red.
(d) both are red, given that at least one is red.

2-J   Problem 2-D gave the joint distribution of random variables R and G:

		\(G\):				
		0	1	2	3	
	0	1/30	6/30	6/30	1/30	14/30
R:	1	3/30	8/30	3/30	0	14/30
	2	1/30	1/30	0	0	2/30
		5/30	15/30	9/30	1/30	1

Find the conditional p.f. of G given $R = 1$, and the conditional p.f. of R given $G = 1$.

Section 2.4

2-K A laboratory test for detecting a certain disease occasionally gives a false positive or a false negative. Suppose 3% of healthy individuals get a false positive: $P(+ \mid H) = .03$, and 2% of diseased individuals get a false negative: $P(- \mid D) = .02$. Given that 1 person in 1000 has the disease: $P(D) = .001$, find $P(D \mid +)$, the probability that a person who gets a positive reading has the disease.

2-L Suppose that in a certain high school, the sizes of the sophomore, junior, and senior classes are equal. Suppose further that 10% of the sophomores, 15% of the juniors, and 20% of the seniors smoke.
(a) Find the proportion of smokers in the high school.
(b) Find the proportion of sophomores among the smokers in the school.
(c) Find the probability that a student seen smoking is a junior.

2-M A contestant in the TV show "Let's Make a Deal" was presented with 3 boxes. One box contained the key to a Lincoln Continental. She chose box number 2. The MC then opened box number 1 and showed that it was empty. He suggested that her chances had now gone up from 1 in 3 to 1 in 2! Is this correct? Assume that the MC knew which box contained the key and would not spoil the fun by opening that box.

Sections 2.5-2.6

2-N Suppose $P(E) = .3$ and $P(F) = .5$. Find $P(EF)$ and $P(E \cup F)$ when
(a) E and F are *disjoint*.   (b) E and F are *independent*.

2-O A random variable X has two possible "values": $\{r, s\}$. A second random variable Y has three possible values: $\{a, b, c\}$. Their joint distribution is defined by the accompanying table of probabilities.

	a	b	c
r	1/4	1/8	1/8
s	1/4	0	1/4

(a) Are events $[X = r]$ and $[Y = a]$ independent?
(b) Are the variables X and Y independent?
(c) Find $P(X = s \mid Y = c)$.
(d) Construct a probability table with the same marginal distributions but which makes X and Y independent variables.

2-P Construct the model for the throw of three dice in which the throws are independent, and find the probability that the total number of points is 5.

2-Q Show that if A, B, and C are independent events, then A is independent of the event $B \cap C$.

2-R   Thirteen cards are dealt from the 52 cards in a standard deck, one at a time (after thorough shuffling). Find the probability
(a) that the 8th card dealt is a heart.
(b) that the 8th card dealt is a heart, given that the last two dealt are hearts.
(c) that the 5th and 6th cards dealt are both hearts.
(d) that the 8th card dealt is a heart, given that it is an ace.

2-S   A bowl contains 5 white and 3 black chips. Two chips are selected at random, one at a time. Find the probability that the first is black, given that the second is white,
(a) if there is replacement and mixing after the first selection.
(b) if there is no replacement.

Chapter 2: Solutions

2-A   **Solution:**

(a) The complement of the event $[X > 2]$ is the event $[X = 1$ or $2]$:

$$P(X > 2) = 1 - f(0) - f(1) = 1 - 1/2 - 1/4 = 3/4.$$

(b) $P(X = 1,$ or 3, or 5, or ...$) = 1/2 + 1/8 + 1/32 + \cdots = \dfrac{1/2}{1 - 1/4} = \dfrac{2}{3}.$

[The infinite series is geometric with first term 1/2 (factor it out) and common ratio 1/4.]

Another way to approach this is to observe that X is odd if *either* we get heads on the first toss, *or* we get two tails and then it takes an odd number of additional tosses to get the first heads:

$$p = P(X \text{ is odd}) = \tfrac{1}{2} + \tfrac{1}{4}p.$$

Solving for p yields $p = 2/3$.

2-B   **Solution:**

"Random selection" means that the $\binom{52}{2} = 1326$ combinations of two cards are equally likely. So to find the probability, we count the combinations of two cards that include 0 hearts, with 1 hearts, and with 2 hearts: $\binom{13}{0}\binom{39}{2} = 741$, $\binom{13}{1}\binom{39}{1} = 507$, and $\binom{13}{2}\binom{39}{0} = 78$, respectively. The p.f. is

$$f(0) = \frac{741}{1326} = \frac{57}{102}, \quad f(1) = \frac{507}{1326} = \frac{39}{102}, \quad f(2) = \frac{78}{1326} = \frac{6}{102}.$$

2-C   **Solution:**

Table 1 (below) shows the 100 possible pairs as cells in a two-way table, with the probability .01 inserted in each cell.

	i									
j	0	1	2	3	4	5	6	7	8	9
0	.01	.01	.01	.01	.01	.01	.01	.01	.01	.01
1	.01	.01	.01	.01	.01	.01	.01	.01	.01	.01
2	.01	.01	.01	.01	.01	.01	.01	.01	.01	.01
3	.01	.01	.01	.01	.01	.01	.01	.01	.01	.01
4	.01	.01	.01	.01	.01	.01	.01	.01	.01	.01
5	.01	.01	.01	.01	.01	.01	.01	.01	.01	.01
6	.01	.01	.01	.01	.01	.01	.01	.01	.01	.01
7	.01	.01	.01	.01	.01	.01	.01	.01	.01	.01
8	.01	.01	.01	.01	.01	.01	.01	.01	.01	.01
9	.01	.01	.01	.01	.01	.01	.01	.01	.01	.01

Table I

The pairs with a given sum $i + j = c$ lie on a diagonal line, such as shown for the sum 13 by circling the 6 pairs with this sum: $(4, 9)$, $(5, 8)$, $(6, 7)$, $(7, 6)$, $(8, 5)$, $(9, 4)$. Thus, $f_Y(13) = .06$. The sums giving $f(k)$ for $k = 0$, 1, ... start with .01 for the pair $(0, 0)$, .02 for the two pairs $(0, 1)$ and $(1, 0)$, increasing in steps of .01 until you get to the longest diagonal (where the sums are 9), and then decreasing in steps of .01 back down to .01 for $f(18)$. The result can be expressed in a formula:

$$f_Y(k) = \frac{10 - |k - 9|}{100}, \quad k = 0, 1, ..., 18.$$

The bar graph representing this p.f. has a triangular outline.

Similarly, differences $i - j$ are constant along diagonals in the other direction. The constant on the three circled is $7 - 0 = 8 - 1 = 9 - 2 = 7$. Then, moving the diagonal across the array, we again find that the distribution has a triangular shape, this time centered at 0 (the value of the difference along the longest diagonal, where $i = j$). If you like formulas, we can write one for the p.f. of Z:

$$f_Z(k) = \frac{10 - |k|}{100}, \quad k = -9, -8, ..., +9.$$

2-D   Solution:
(a) The possible values of G are 0, 1, 2, 3, and the possible values of R are 0, 1, 2. But not all of the 4×3 pairs (i, j) are possible. Some sample calculations:

$$f(1, 0) = P(R = 1, G = 0) = \frac{\binom{2}{1}\binom{4}{0}\binom{4}{2}}{\binom{10}{3}} = \frac{3}{30},$$

$$f(2, 1) = P(R = 2, G = 1) = \frac{\binom{2}{2}\binom{4}{1}\binom{4}{0}}{\binom{10}{3}} = \frac{1}{30},$$

$$f(1, 1) = P(R = 1, G = 1) = \frac{\binom{2}{1}\binom{4}{1}\binom{4}{1}}{\binom{10}{3}} = \frac{8}{30}.$$

The complete table of joint probabilities is shown below.

(b) The marginal probabilities are given in the right and lower margins of the table.

		G: 0	1	2	3	
	0	1/30	6/30	6/30	1/30	14/30
R:	1	3/30	8/30	3/30	0	14/30
	2	1/30	1/30	0	0	2/30
		5/30	15/30	9/30	1/30	1

(c) $R + G = 2$ in the pairs $(0, 2)$, $(1, 1)$, $(2, 0)$, with total probability $(1 + 8 + 6)/30 = 1/2$.

(d) $R = G$ along the "main diagonal"—the pairs $(0, 0)$, $(1, 1)$, $(2, 2)$, with total probability $(1 + 8 + 0)/30 = 3/10$.

2-E Solution:
(a) *All* multiples of 4 are even, so the answer is 1.

(b) There are 5 even digits, and 2 are multiples of 4: $2/5$.

(c) There are three multiples of 3: 3, 6, 9, of which 1 is even: $1/3$.

(d) Among the 5 even numbers, only the 6 is a multiple of 3: $1/5$.

(e), (f) $TF = \emptyset$, so both $P(T \mid F)$ and $P(F \mid T)$ are 0.

2-F Solution:
The proportions are interpreted as probabilities: $P(\text{age} > 20) = .947$, and $P(\text{age} > 65) = .499$. Since the event [age > 65] is a subset of [age > 20], their intersection is [age > 65], and
$$P(\text{age} > 65 \mid \text{age} > 20) = \frac{P(\text{age} > 65)}{P(\text{age} > 20)} = \frac{.499}{.947} \doteq .527.$$

2-G Solution:
The percentages given are *conditional*, given the sex. We need to know the proportion of males and females:

(a) $P(\text{male}) = 18595/25683 = .724$, and so $P(\text{female}) = .276$.

(b) $P(\text{poison}) = P(\text{poison, male}) + P(\text{poison, female})$
$$= P(p \mid m)P(m) + P(p \mid f)P(f) = .158 \times .724 + .43 \times .276 = .233.$$

(c) $P(f \mid p) = \dfrac{P(f \text{ and } p)}{P(p)} = \dfrac{.43 \times .276}{.233} = .509$

(d) $P(m \mid e) = \dfrac{P(m \text{ and } e)}{P(e)} = \dfrac{.635 \times .724}{.635 \times .724 + .357 \times .276} = .8235.$

2-H Solution:
Parts (a) and (b) deal with *order,* so the appropriate sample space is one of sequences; however, the model is defined by the statement that the items are drawn one at a time at random—equally likely outcomes, and the probability (based on equally likely outcomes) for the second selection is a conditional probability. The questions in parts (c)-(e) do not involve order, and we can use the sample space of combinations, if we recall that when selections are done one at a time as described, the possible combinations are equally likely.

(a) $P(DG) = P(\text{1st D}) \cdot P(\text{2nd G} \mid \text{1st D}) = \frac{2}{10} \cdot \frac{8}{9} = \frac{8}{45}.$

[Since the $10 \cdot 9$ sequences are equally likely, we can count the number in which the first is D and the second G (given that the first is D): $\frac{2 \cdot 8}{10 \cdot 9}.$]

(b) Same reasoning and answer as in (a): $\frac{8}{10} \cdot \frac{2}{9}.$

(c) [DG or GD] = [DG] $\cup$ [GD], so add answers to (a) and (b): $\frac{16}{45}.$

(d) Using combinations, $\frac{\binom{2}{2}}{\binom{10}{2}} = \frac{1}{45} = P(DD) = P(\text{1st D})P(\text{2nd D} \mid \text{1st D}).$

(e) As in (d), $\frac{\binom{8}{2}}{\binom{10}{2}}$ or just subtract (c) and (d) from 1: $1 - \frac{17}{45}.$

2-I   **Solution:**

(a) We need look only at the set of red cards; and the question does not involve order, so we consider combinations: There are $\binom{26}{2} = 325$ possible red pairs, of which $\binom{13}{2} = 78$ are pairs of hearts. Answer: $78/325 = .24.$

(b) There are 13 cards in any one suit, so $\binom{13}{2} = 78$ possible pairs; and there are just 3 face cards—3 different pairs: $3/78 = 1/26.$

(c) $P(\text{both red} \mid \text{2nd red}) = \dfrac{P(\text{both red})}{P(\text{2nd R})} = \dfrac{\binom{26}{2} \big/ \binom{52}{2}}{1/2} = \dfrac{25}{51}.$

(d) $P(\text{both red} \mid \text{not both black}) = \dfrac{P(\text{both red})}{P(\text{not both black})} = \dfrac{\binom{26}{2} \big/ \binom{52}{2}}{1 - \binom{26}{2} \big/ \binom{52}{2}} = \dfrac{25}{77}.$

2-J   **Solution:**

The table of joint probabilities is repeated below, with probabilities outlined that are relevant to the two conditions given.

When R is known to be 1, G has the values 0, 1, 2 with odds 3:8:3, which convert (upon division by the total) to probabilities 3/14, 8/14, 3/14. When G is known to have the value 1, R has the values 0, 1, 2 with odds 6:8:1, which convert to probabilities 6/15, 8/15, 1/15.

G:

	0	1	2	3	
0	1/30	6/30	6/30	1/30	14/30
R: 1	3/30	8/30	3/30	0	14/30
2	1/30	1/30	0	0	2/30
	5/30	15/30	9/30	1/30	1

2-K   **Solution:**

We want $P(D \mid +)$ but are given $P(+ \mid D)$. This is the kind of situation that calls for Bayes' theorem:

$$P(D \mid +) = \frac{P(+ \mid D)P(D)}{P(+ \mid D)P(D) + P(+ \mid H)(P(H)}.$$

[See (3) in §2.4.] Each of the probabilities on the right is given or implied in the statement of the problem:

$$P(D \mid +) = \frac{.98 \times .001}{.98 \times .001 + .03 \times .999} = .0317.$$

[You should construct a tree diagram: at the first node, split into D and H (probabilities .001 and .999); at the second node, split into $+$ and $-$.] Why is this probability so small, when the test equipment seems to be fairly accurate? Essentially, it's because the disease is rare. It is more than 29 times as likely that the reading is a false positive than that a diseased person has come along. But the positive reading increases the probability that a person has a disease from .001 to .0317—a 30-plus fold increase.

2-L   Solution:
(a) The law of total probability [(2) in §1.7] tells us that

$$P(S) = .10 \times 1/3 + .15 \times 1/3 + .20 \times 1/3 = .15.$$

(b) Bayes' theorem says that the proportions of sophomores, juniors, and seniors among the smokers are proportional to the terms in the sum that gives us $P(S)$ in (a). So the odds are 10:15:20, or 2:3:4; we make these into probabilities (or population proportions) by dividing the $2 + 3 + 4$ or 9: 2/9, 3/9, 4/9. The proportion of sophomores is 2/9.

(c) By the same token, the proportion of juniors is 3/9. This fraction is the probability that the student is a junior, given that he/she smokes.

2-M   Solution:
To answer the question, we need to know how the MC would pick a box to open if the key were in the box the contestant had picked. (If she did *not* have it already, it would be in one of the unopened boxes, and he would open the one without the key.) We assume that he chooses at random. Thus, with these assumptions,

$$P(\text{shows 1} \mid \text{key in 1}) = 0,$$
$$P(\text{shows 1} \mid \text{key in 2}) = 1/2,$$
$$P(\text{shows 1} \mid \text{key in 3}) = 1.$$

Then,

$$P(\text{key in 2} \mid \text{shows 1}) = \frac{P(\text{shows 1} \mid \text{key in 2}) \cdot P(\text{key in 2})}{P(\text{shows 1})}$$

$$= \frac{\frac{1}{2} \times \frac{1}{3}}{0 \times \frac{1}{3} + \frac{1}{2} \times \frac{1}{3} + 1 \times \frac{1}{3}} = \frac{1}{3}.$$

(Again we've used the law of total probability for the denominator.) This says that the MC has given no information in opening a box you know that he knows is empty. The MC (Monte Hall) later put it this way in a letter to a statistician. "Oh, and incidentally, after one [box] is seen to be empty, her chances are no longer 50-50 but remain what they were in the first place, one out of three. It just seems to the contestant that one box having been eliminated, she stands a better chance. Not so." But what actually happened was that the contestant said to him (when he suggested a change in odds), "I don't know about that, but I'll trade my box for the unopened one." (She was smarter than he thought: If each box originally had 1 chance in 3 of containing the key, her box had 1 chance in 3 and the MC's had 2 chances in 3; when he opened the one box, the 2 chances in 3 that the key was in his boxes now resided in his unopened box. A trade is a wise move.)

2-N   Solution:

(a) Saying that E and F are disjoint means that $EF = \emptyset$, so $P(EF) = 0$. And then $P(E \cup F) = P(E) + P(F) - P(EF) = .3 + .5 - 0 = .8$.

(b) If E and F are independent, then $P(EF) = P(E)P(F) = .3 \times .5 = .15$, and $P(E) + P(F) - P(EF) = .3 + .5 - .15 = .65$.

2-O   Solution:

(a) Yes, because $P(X = r)P(Y = a) = \frac{1}{2} \cdot \frac{1}{2} = \frac{1}{4} = P(X = r, Y = a)$.

(b) No, because not *all* table entries are the products of corresponding marginal probabilities; in particular the 0 for the pair (s, b) is not the product of the marginal totals, 1/8 and 1/2.

(c) When $Y = c$, $X = s$ is twice as likely as $X = r$: $P(X = s \mid Y = c) = \frac{2}{3}$.

(d) Just multiply the marginals: $P(i, j) = P(i)P(j)$ for each pair (i, j):

	a	b	c	
r	1/4	1/16	3/16	1/2
s	1/4	1/16	3/16	1/2
	1/2	1/8	3/8	

(Observe that this process creates an array in which rows are proportional and columns are proportional.)

2-P   **Solution:**
The outcomes are triples (i, j, k) in which each element is a number taken at random from the integers 1 to 6. To incorporate independence into the model, we define $P(i, j, k) = P(i)P(j)P(k) = (1/6)^3$. So the 216 triples are equally likely, and we count the triples in which the sum is 5. They are: 113, 131, 311, 122, 212, 221—six in number, so $P(5) = 6/216$.

2-Q   **Solution:**
Independence of the three events means (i) $P(ABC) = P(A)P(B)P(C)$, and (ii) pairwise independence. To show independence of A and BC, we show $P[A(BC)] = P(A)P(BC)$. First, by (ii), $P(BC) = P(B)P(C)$. Then

$$P[A(BC)] = P(ABC) = P(A)[P(B)P(C)] = P(A)P(BC).$$

2-R   **Solution:**
In sampling without replacement from a finite population, the outcomes of the various selections are exchangeable, as are all pairs of selections.
(a) This equals the probability that the 1st is a heart: 1/4.

(b) This equals the probability that the 3rd is a heart, given that the 1st and 2nd are hearts: With 2 hearts gone, 11 remain: 11/50.

(c) This equals the probability that the first two are hearts: $\frac{13}{52} \cdot \frac{12}{51}$.

(d) The variable "suit" is independent of the variable "denomination," so the information that the card is an ace is irrelevant: $P(\text{8th is heart}) = 1/4$.

2-S   **Solution:**
(a) With replacement and mixing, we can consider the selections independent, and $P(\text{1st black}) = 3/8$.

(b) The selections are exchangeable, so (with 1 white chosen first),

$$P(\text{1st B} \mid \text{2nd W}) = P(\text{2nd B} \mid \text{1st W}) = 3/7.$$

Not noticing the exchangeability, we could use Bayes' theorem: First find

$$P(\text{2nd W}) = P(\text{2nd W} \mid \text{1st B})P(\text{1st B}) + P(\text{2nd W} \mid \text{1st W})P(\text{1st W})$$

$$= \frac{5}{7} \cdot \frac{3}{8} + \frac{4}{7} \cdot \frac{5}{8} = \frac{15 + 20}{56} = \frac{5}{8} = P(\text{1st W}).$$

$$P(\text{1st B} \mid \text{2nd W}) = \frac{P(\text{2nd W} \mid \text{1st B})P(\text{1st B})}{P(\text{2nd W})} = \frac{15}{15 + 20} = \frac{15}{35} = \frac{3}{7}.$$

CHAPTER 3: Additional Problems

Sections 3.1-3.2

3-A   In Problem 2-D we found the distribution of the number of green socks in a random selection of 3 from a drawer with 10 socks, 4 of them green: The possible values are 0, 1, 2, 3, and the corresponding probabilities are 5/30, 15/30, 9/30, 1/30. Find the mean number.

3-B   Three numbers are selected at random without replacement from the integers 1 through 9. Let X denote the largest of the three numbers, and find EX.

3-C   A children's game includes a spinner with six equal sectors, marked -5, 1, 2, 3, 4, 5. After a spin, the player's token advances a corresponding number of spaces on the game board.
(a) Find EX, where X is the number of spaces advanced in one turn.
(b) Find the mean number of spaces advanced in six turns.

3-D   Let S denote the sum of two random digits—two numbers selected from the digits 0, 1, ..., 9 at random and independently. In Problem 2-C we found the p.f. for S to be
$$f(k) = \frac{10 - |k - 9|}{100}, \quad \text{for } k = 0, 1, ..., 18.$$
Find ES.

3-E   On the sample space of the preceding problem (pairs of random digits), let Z denote the difference $i - j$ (first digit minus the second).
(a) Find EZ.   (b) Find $E(Z^2)$.

Section 3.3

3-F   In Problem 2-D we found the distribution of the number of green socks in a random selection of 3 from a drawer with 10 socks, 4 of them green: The possible values are 0, 1, 2, 3, and the corresponding probabilities are 5/30, 15/30, 9/10, 1/30. The mean, from Problem 3-A, is 1.2; find the standard deviation of the number of green socks.

3-G For the distribution in the preceding problem, find the mean deviation.

3-H In Problem 3-E we found the mean and mean square of the difference in two (independent) random digits. Find the variance and s.d.

3-I Let X denote the number of red cards in a bridge hand (13 cards) and Y, the number of black cards. Show that X and Y have the same variance, and that the variance of the sum is not the sum of the variances.

Section 3.4-3.5

3-J Problem 2-D gives the joint distribution of variables G and R, as a table of joint probabilities:

G:

	0	1	2	3	
0	1/30	6/30	6/30	1/30	14/30
R: 1	3/30	8/30	3/30	0	14/30
2	1/30	1/30	0	0	2/30
	5/30	15/30	9/30	1/30	1

(a) Find the means, variances, and covariance of R and G.
(b) Find the correlation coefficient.

3-K Verify the formula for the variance of a sum [(3) of §3.5] in the case of the distribution in the preceding problem:
(a) Find the distribution of $R + G$.
(b) Using (a), find $\text{var}(R + G)$.
(c) Find $\text{var } R + \text{var } G + 2\,\text{cov}(R, G)$ using values from Problem 3-J.

3-L Find the coefficient of correlation of the pair (X, Y) with distribution given by the accompanying table of joint probabilities.

Y:

		1	2	3
	1	0	0	1/2
X:	2	0	1/3	0
	3	1/6	0	0

3-M Find $\text{var}(X + Y + Z)$, when $\sigma_X^2 = 1$, $\sigma_Y^2 = 4$, $\sigma_Z^2 = 9$, and
(a) X, Y, and Z are independent.
(b) $\sigma_{X,Y} = -1$, $\sigma_{X,Z} = 2$, $\sigma_{Y,Z} = 0$.

3-N Given $\sigma_X^2 = 1$, $\sigma_Y^2 = 4$, $\sigma_{X,Y} = -1$, find the following:
(a) $\text{var}(2X - 3Y + 5)$.   (b) $\text{cov}(X - 2Y, 3X + 4Y)$.

3-O   Find the variance of the sum of 10 independent, random digits.

3-P   Find the mean and variance of the number of points in a hand of 13 cards dealt from a standard deck. (The number of points assigned to a card is given in Problem 2-34 of the text.)

Section 3.6

3-Q   The distribution of X is characterized by its p.g.f., $\eta(t) = (3 + 4t + 2t^2)/9$. Consider 3 independent trials, yielding $Y = X_1 + X_2 + X_3$. Find
(a) the p.f. of X.   (b) $\eta_Y(t)$.   (c) $P(Y = 3)$.

3-R   Find the probability of getting a total of 7 with the throw of four dice.

Chapter 3: Solutions

3-A   **Solution:**
The mean is the sum of the products: $\; 0 \cdot \frac{5}{30} + 1 \cdot \frac{15}{30} + 2 \cdot \frac{9}{30} + 3 \cdot \frac{1}{30} = 1.2.$

3-B   **Solution:**
To find the mean, we first need the p.f. of X, $f(k) = P(\text{largest is } k)$.
The largest will be k if and only if the two smaller integers are chosen
1, 2, ..., $k-1$:

$$ f(k) = \frac{\binom{k-1}{2}}{\binom{9}{3}}, \quad k = 3, 4, ..., 9. $$

The mean is then

$$ \sum_{3}^{9} k \frac{\binom{k-1}{2}}{\binom{9}{3}} = \frac{3 \times 1 + 4 \times 3 + 5 \times 6 + 6 \times 10 + 7 \times 15 + 8 \times 21 + 9 \times 28}{84}, $$

which is 7.5.

Another way to get the p.f. is to find the c.d.f.:

$$ P(X \le k) = P(\text{all three come from 1, 2, ..., } k) = \frac{\binom{k}{3}}{\binom{9}{3}} = \frac{k(k-1)(k-2)}{504}, $$

and take differences: $\; P(X = k) = P(X \le k) - P(X \le k - 1)$.

3-C   **Solution:**
(a) If we assume the sectors to be equally likely, their probabilities are
each 1/6, and the mean value is just the sum of the possible values divided
by 6: $\; EX = 10/6$.

(b) The number advanced in six spins is the sum of the numbers advanced
in the individual spins. The mean of the sum is the sum of the means, or
$6 \times 10/6 = 10$.

3-D   **Solution:**
There at least three ways one can do this.
(i) The long way is to add the products $kf(k)$:

$$ ES = 0 \times .01 + 1 \times .02 + \cdots + 18 \times .01 = 9. $$

(ii) A shorter way is to observe that the sum can be thought of as the
sum of the random variables X, the first digit, and Y, the second digit.
Both X and Y are uniform (equally likely values) on 0, 1, ..., 9, with
mean 4.5, the midpoint of their distribution. Then $ES = EX + EY = 9$.

(iii) Easier still, we simply note that the given p.f. of S is symmetric
about the value 9: $\; f(9 - k) = f(9 + k)$, so $ES = 9$.

3-E   Solution:

(a) The easy way is to write $Z = X - Y$ (X and Y from 3-D), and obtain $EZ = EX - EY = 0$, since X and Y are identically distributed. Another way is to note that the p.f., which we found in Problem 2-C, is

$$f_Z(k) = \frac{10 - |k|}{100}, \quad k = -9, -8, ..., +9.$$

This is symmetric about 0, which is therefore the mean.

(b) At this point, there is no quick way. All we can do is square each of the possible differences and multiply by the probability weighting; there will be two numerators of the form

$$9^2 \times 1 + 8^2 \times 2 + 7^2 \times 3 + 6^2 \times 4 + 5^2 \times 5 + 4^2 \times 6 + 3^2 \times 7 + 2^2 \times 8 + 1^2 \times 9,$$

equal to 825, and

$$E(Z^2) = \frac{825 + 825}{100} = 16.5.$$

3-F   Solution:

The average square is $0^2 \cdot \frac{5}{30} + 1^2 \cdot \frac{15}{30} + 2^2 \cdot \frac{9}{30} + 3^2 \cdot \frac{1}{30} = 2$, so the variance is $2 - 1.2^2 = .56$, and $\sigma = \sqrt{.56} = .748$. (You should sketch the p.f., mark the mean and the s.d., to check that the values found appear reasonable.)

3-G   Solution:

The deviations from the mean (1.2) are -1.2, $-.2$, $.8$, 1.8, with absolute values 1.2, .2, .8, 1.8. These are weighted with probabilities of 0, 1, 2, 3:

$$\text{m.a.d.} = 1.2 \times \frac{5}{30} + .2 \times \frac{15}{30} + .8 \times \frac{9}{30} + 1.8 \times \frac{1}{30} = \frac{18}{30} = .60,$$

slightly less (as we should expect) than the s.d.

3-H   Solution:

The variance is the mean square minus the square of the mean. In this instance, the mean is 0, so the variance is the mean square: 16.5. And s.d. is the square root, $\sigma = 4.062$. (Again, a graph of the p.f. with the mean and s.d. marked would be educational.)

3-I   Solution:

The variables X and Y are exchangeable, because there are as many red cards in the deck as there are black, and the p.f. is a symmetric function of its arguments:

$$f(x, y) = \frac{\binom{26}{x}\binom{26}{y}}{\binom{52}{13}}, \quad x + y = 13, \quad x > 0, \, y > 0.$$

Thus, var $X =$ var $Y > 0$. But $X + Y = 13$, every time you deal—no variability, so $\text{var}(X + Y) = 0$.

3-J **Solution:**

(a) In Problems 3-A, 3-F we found $\mu_G = 1.2$ and $\sigma_G^2 = .56$. From the marginal probabilities for R (above) we have $ER = 18/30 = .6$, and $E(R^2) = 22/30$, so $\sigma_R^2 = 22/30 - (3/5)^2 = 28/75$.

To get the covariance, we first find $E(RG)$. The product rg for the first row and column is 0, so there are only 3 nonzero terms:

$$E(RG) = 1 \times \frac{8}{30} + 2 \times \frac{3}{30} + 2 \times \frac{1}{30} = \frac{8}{15}.$$

Then

$$\sigma_{R,G} = \frac{8}{15} - \frac{6}{5} \cdot \frac{3}{5} = -\frac{14}{75}.$$

(b) $\rho = \dfrac{\sigma_{R,G}}{\sigma_G \sigma_R} = -.408.$

3-K **Solution:**

(a) Using the table in 3-J, we see that the possible values of $R + G$ are 0, 1, 2, 3. Corresponding probabilities are obtained by summing probabilities along the diagonals where sums are constant: 1/30, 9/30, 15/30, and 5/30.

(b) Comparing this distribution with that of G, we see that the variances must be the same: $\text{var}(R+G) = .56 = 14/25$.

(c) $\sigma_G^2 + \sigma_R^2 + 2\sigma_{G,r} = \dfrac{14}{25} + \dfrac{28}{75} + 2 \times \dfrac{-14}{75} = \dfrac{14}{25} = $ (b).

3-L **Solution:**

The means are $\mu_X = \frac{1}{2} + \frac{2}{3} + \frac{3}{6} = \frac{5}{3}$, and $\mu_Y = \frac{1}{6} + \frac{2}{3} + \frac{3}{2} = \frac{7}{3}$. The variances are equal because the distribution pattern of X is the same as for Y, but reversed:

$$\text{var } Y = \text{var } X = \frac{1}{2} + \frac{4}{3} + \frac{9}{6} - \left(\frac{5}{3}\right)^2 = \frac{5}{9}.$$

The covariance is

$$\sigma_{X,Y} = \sum x_i y_j f(x_i, y_j) - \mu_X \mu_Y = \frac{3}{6} + \frac{4}{3} + \frac{3}{2} - \frac{5}{3} \cdot \frac{7}{3} = -\frac{5}{9},$$

so

$$\rho = \frac{-5/9}{\sqrt{5/9}\sqrt{5/9}} = -1.$$

This could have been predicted: All of the probability was assigned to points *on the line* $x + y = 4$ (with negative slope).

3-M **Solution:**

(a) When the variables are independent, the variance is additive, so the variance of the sum is $1 + 4 + 9 = 14$.

(b) The variance of the sum is the sum of the variances plus twice the sum of all the covariances: $\text{var}(X + Y + Z) = 14 + 2 \times (-1) + 2 \times 2 = 16$.

3-N   Solution:

(a) Since $\text{var}(X+c) = \text{var}\,X$ for any constant c, we can ignore the 5:

$$\text{var}(2X - 3Y + 5) = \text{var}(2X) + \text{var}(-3Y) + 2\,\text{cov}(2X, -3Y).$$

In variances, constant multiples are removed as squares; in covariances, constant factors can be taken out as is:

$$\text{var}(2X - 3Y + 5) = 4\sigma_X^2 + 9\sigma_Y^2 - 12\sigma_{X,Y} = 52.$$

(b) To find the covariance of two linear combinations, we find the covariance of each term in one with each term in the other, and sum:

$$\text{cov}(X - 2Y, 3X + 4Y) = \text{cov}(X, 3X) + \text{cov}(X, 4Y)$$
$$+ \text{cov}(-2Y, 3X) + \text{cov}(-2Y, 4Y).$$

Again moving constants outside the "cov," and using $\sigma_{X,Y} = \sigma_{Y,X}$ and $\sigma_{X,X} = \sigma_X^2$, we obtain

$$\text{cov}(X - 2Y, 3X + 4Y) = 3\sigma_X^2 + 4\sigma_Y^2 + 4\sigma_{X,Y} - 6\sigma_{Y,X} = 21.$$

3-O   Solution:

For a single random digit (chosen at random from $0, 1, \ldots, 9$), the mean value is 4.5, the balance point. The average square is

$$EX^2 = (0^2 + 1^2 + 2^2 + \cdots + 9^2)/10 = 28.5.$$

So $\text{var}\,X = 28.5 - 4.5^2 = 8.25$—for each of the random digits. Because the digits are assumed independent, the variance of their sum is the sum of their variances: $10 \times 8.25 = 82.5$.

3-P   Solution:

Let Y be the total number of points assigned to the 13 cards. This is the sum of the numbers assigned to the individual cards: $X_1 + \cdots + X_{13}$. In Problems 3-4 and 3-15, we found the mean and variance of each X to be $\mu = 10/13$, and $\sigma^2 = 290/169$. The mean of the sum is the sum of the means: $EY = 13\mu_X = 10$.

The variables X_i, although exchangeable, are not uncorrelated, so the variance of Y will involve covariances: $\sigma_{ij} = \text{cov}(X_i, X_j)$. Since the variables are exchangeable, these covariances are all the same, and we could find the common value by finding σ_{12}. With this, we have

$$\text{var}\,Y = 13\sigma^2 + 13 \times 12 \times \sigma_{12}.$$

It is not difficult to find σ_{12}, but the easy way is to use the trick leading up to (9) in §3.5, where we found $\sigma_{12} = -\sigma^2/(N-1)$. Here, $N = 52$, so $\sigma_{12} = -\sigma^2/51$, and $\text{var}\,Y = 13\sigma^2\left(1 - \frac{12}{51}\right) = \frac{290}{17}$.

3-Q   **Solution:**

(a) The possible values are the exponents of the powers of t: 0, 1, 2, and their probabilities are the corresponding coefficients: $f(0) = 3/9$, $f(1) = 4/9$, and $f(2) = 2/9$.

(b) The p.g.f. of Y is the 3rd power of the p.g.f. of X: $(3 + 4t + 2t^2)^3/9^3$.

(c) The square of the numerator of η_Y is: $9 + 24t + 28t^2 + 16t^3 + 4t^4$. To get the cube, we multiply this by η_Y: $\quad\underline{\times \ (3 + 4t + 2t^2)}$

The product has three terms in t^3: $(3 \times 16 + 4 \times 28 + 2 \times 24)t^3$. The coefficient, divided by 9^3, is the desired probability: $208/729$.

3-R   **Solution:**

One way to do this is to enumerate the ways of getting a total of 7 and divide this count by 6^4. Using the p.g.f. is more systematic and less apt to overlook some way; for a single die,

$$6\eta(t) = t + t^2 + t^3 + t^4 + t^5 + t^6 = \frac{t(1 - t^6)}{1 - t}.$$

The p.g.f. of the total is the 4th power of the p.g.f. for a single die:

$$1296\eta^4(t) = t^4(1 - t^6)^4(1 - t)^{-4}.$$

In this product, we use the binomial theorem for the second factor, so that

$$t^4(1 - t^6)^4 = t^4 - 4t^{10} + 6t^{16} - \cdots.$$

For the third factor, we use the negative binomial expansion:

$$(1 - t)^{-4} = 1 + 4t + 10t^2 + 20t^3 + \cdots.$$

In the product of these two series, there is only one term with t to the 7th power: $20t^7$, so the desired probability is $20/1296$, or about .015.

CHAPTER 4 Additional Problems

Sections 4.1-4.2

4-A Find the probability that in four tosses of a fair die, the side "3" will turn up at most once.

4-B Each of 12 people tastes two colas, one in a glass marked Q and the other in a glass marked M. Each person is asked to pick the Pepsi. Suppose both glasses actually contain Pepsi. Let Y denote the number of people who choose the glass marked M.
(a) What is the expected value of Y?
(b) What is the probability that 9 or more of the 12 choose glass M, if the label has no effect on which glass they choose?
[Pepsi conducted such an experiment some years ago, to show that in a result advertised by Coke, people really were expressing a preference for the letter M over the letter Q.]

4-C In Problems 2-32 and 2-33 we regarded a baseball player's successive times at bat (AB's) as independent trials of the same Bernoulli experiment, in which p, probability of a hit, is the player's batting average. With this assumption, find the probability that "300 hitter" (one whose p is .300) gets at least one hit in four AB's.

4-D Find the probability that the batter in Problem 4-C gets *exactly* one hit in a particular game (in which he has 4 AB's), given that he gets at least one hit.

4-E Find the value of the following sum: $\sum_{k=0}^{15} k^2 \binom{15}{k} .4^k .6^{15-k}$.

Sections 4.3-4.4

4-F An instructor assigned 7 possible essay topics, stating that 3 would be selected for the final exam. One student studied 5 of the 7 topics thoroughly and ignored the others. Assuming a random selection of topics on the exam,
(a) give the probability distribution of Y, the number of questions for which the student was prepared.
(b) find the mean and variance of the number in (a).

4-G   A group of 12 patients—6 men and 6 women—are to be used in evaluation of a drug. Some will be assigned the drug therapy and the others (as a "control") are to be given a placebo.
(a) Suppose 6 are selected at random to be assigned the drug. Let Y denote the number of men in this treatment group. Calculate $P(Y = 3)$, the probability that there are 3 men and 3 women in each group.
(b) Suppose as each patient is considered, he or she is assigned to either the treatment group or to the control group by the toss of a coin. Find the probability that there will be 6 in each group.
(c) Find the probability that, with the procedure given in (b), there are exactly 3 men and 3 women in the treatment group.

4-H   A bag containing 50 jelly beans includes exactly 8 that are black. We take one at a time from the bag (without looking in the bag), and keep on until we get a black one. Find the probability that it takes 5 tries to get 2 black ones.

4-I   Suppose one selects jelly beans one at a time, as in the Problem 4-H, but from a very large container (large enough that the supply can be thought of as infinite), with the proportion of blacks the same: $p = .16$. Now
(a) what is the probability that it takes exactly 5 tries to get 2 blacks?
(b) what is the expected number of tries to get blacks?

4-J   Compare the following hypergeometric probability with its binomial approximation:
$$\frac{\binom{500}{4}\binom{500}{4}}{\binom{1000}{8}}.$$

4-K   In screening donors one at a time to find 4 people with type O blood, find
(a) the average number that must be tested to find the four.
(b) the probability that at most 12 will need to be tested to find four.
Assume that 45% of the population of potential donors are type O.

4-L   Find the expected number of rolls of an ordinary die that would be needed to see each face at least once.

Section 4.5

4-M   Suppose 35% of the viewers in a metropolitan area are watching a certain TV special program. Find the probability that in a random selection of 1,000 viewers, at most 330 of them are watching the program.

4-N   Find the probability of getting exactly 32 heads in 60 tosses of a fair coin.

4-O   Suppose a random sample of 20 is drawn from a population that is 90% white. Find $P(Y \geq 18)$, where Y is the number of whites in the sample.

4-P   Suppose, early in a presidential campaign, only 2% of the voters in a state favor A, while 30% favor B. Consider a random selection of 200 voters.
(a) Find the mean and s.d. of the number in the sample who favor B.
(b) Find the probability that fewer than 50 favor B.
(c) Find the probability that none favor A.
(d) Find the probability that at most three favor A.

Section 4.6

4-Q   Flaws in a manufacturer's carpet occur on the average of one in 50 square yards. Assuming a Poisson distribution, find the probability of
(a) at most one flaw in a 10-square-yard piece.
(b) exactly one flaw in a 10-square-yard piece, given there is at most one.

4-R   An arrival process is assumed to be Poisson, with parameter $\lambda = 24/\text{hr}$.
(a) Find the probability of no more than 10 arrivals in a half-hour period.
(b) Find the probability of at most one arrival in a one-minute period.
(c) Find the probability that the time to the second arrival exceeds 1 min.

4-S   Find the value of $\displaystyle\sum_0^\infty x^2 \cdot \frac{4^x}{x!} e^{-4}$.

4-T   Consider a Poisson process with rate λ, as a model for certain arrivals. Let X and Y denote, respectively, the numbers of arrivals in consecutive unit intervals of time. Find the conditional distribution of X, given that $X + Y = c$, a given constant.

Sections 4.8–4.9

4-U   At one time, plain M&M's came in five colors: brown, orange, yellow, tan and green. The manufacturer claimed the colors were mixed in the proportions 4:2:2:1:1, respectively. A small bag contains 20. Let Y denote the number of yellow's in the bag, etc.
(a) Find the probability that there are four of each color in the bag.
(b) Given $G = 0$, $T = 2$, what is the distribution of (B, O, Y)?
(c) Given $B = 8$, $O = 4$, $Y = 3$, what is the distribution of G?

4-V In a population of 500 people, the distribution according to blood group is as follows: 225 are type O, 200 are type A, 50 are type B, 25 are type AB. We draw a random sample of size 8 without replacement. Let X denote the number of type O, Y the number of type A, and Z the number of type AB in the sample.
(a) Give an exact expression for $P(X = 4)$.
(b) Approximate the probability in (a) based on the fact that the sample size is small compared to 500.
(c) Find the exact probability that $X = 4$, $Y = 2$, and $Z = 1$.
(d) Approximate the answer to (c) using the multinomial distribution.
(e) Use the Poisson p.f. to approximate $P(X + Y + Z \leq 6)$.

4-W Given that X has the f.m.g.f. $.4t + .3t^2 + .2t^3 + .1t^4$,
(a) use this to find the mean and variance of X.
(b) find the p.f. of X.

4-X Let X denote the number of bidding points assigned a card in the system of point count bidding in bridge described in Problems 3-4 and 3-15 of the text and 3-P above. Use the p.g.f. to find the variance of X. [Recall: The possible values are 0, 1, 2, 3, 4, with probabilities 9/13 for 0, and 1/13 for each of the other values.]

Chapter 4: Solutions

4-A   Solution:
The probability of a 3 in a single toss is 1/6. Let Y denote the number of 3's in four tosses. We assume the tosses to be independent. (Without some assumption about their relationship, we can't do anything.) Then Y is binomial—the number of "successes" in a fixed number of independent trials of a Bernoulli experiment. Using the binomial formula, we have

$$P(Y = 0 \text{ or } 1) = \left(\tfrac{5}{6}\right)^4 + \left(\tfrac{4}{1}\right)\left(\tfrac{1}{6}\right)^1\left(\tfrac{5}{6}\right)^3 = \tfrac{125}{144} \doteq .868.$$

4-B   Solution:
(a) With the same drink in both glasses, the probability that the glass marked M is picked is 1/2. Assuming independent trials seems reasonable, in which case Y is binomially distributed: $Y \sim \text{Bin}(12, .5)$. The mean value is $np = 12 \times .5 = 6$.

(b) $P(Y \geq 9) = \sum_{k=9}^{12} f(k \mid .5) = \sum_{k=9}^{12} \binom{12}{k}\left(\tfrac{1}{2}\right)^k\left(\tfrac{1}{2}\right)^{12-k}.$

This sum can be calculated using a hand calculator, but we have a table (Table Ib) that gives tail-probabilities for binomial variables with $n = 12$; it gives .0730 as the desired probability.

4-C   Solution:
This is a binomial probability—the number of hits is $\text{Bin}(4, p)$. But here's an instance in which the complementary event is simpler:

$$P(\text{at least one}) = 1 - P(\text{none}) = 1 - .7^4 \doteq .76.$$

4-D   Solution:
We use the formula that defines the conditional probability of an event:

$$P(\text{exactly } 1 \mid \text{at least } 1) = \frac{P(\text{exactly } 1)}{P(\text{at least } 1)} = \frac{\binom{4}{1}.3 \times .7^3}{1 - .7^4} \doteq .54.$$

(Since the event [exactly 1] is a subset of [at least 1], the probability of the intersection, for the numerator, is the probability of the smaller event.)

4-E   Solution:
The terms in this sum are of the form $k^2 f(k)$, where f is the binomial p.f. for a variable Y which is $\text{Bin}(15, .4)$. The sum is therefore $E(Y^2)$. And

$$E(Y^2) = \text{var } Y + (EY)^2 = npq + (np)^2 = 15 \times .24 + 6^2 = 39.6.$$

4-F **Solution:**

(a) The 7 topics constitute a Bernoulli population, 5 "successes" or 1's, (those for which the student studied), and 2 failures. The number of 1's in a random selection of 3 is hypergeometric: $N = 7$, $M = 5$, $n = 3$. The p.f. of Y is

$$f(k) = \frac{\binom{5}{k}\binom{2}{3-k}}{\binom{7}{3}}, \quad k = 0, 1, 2, 3.$$

(*Either* the name together with parameter values, or the p.f. would be a sufficient answer, when you are asked to "give the distribution.")

(b) The mean is $np = 3 \times 5/7$, and the variance is

$$npq \times \frac{N-n}{N-1} = 3 \times \frac{5}{7} \times \frac{2}{7} \times \frac{7-3}{7-1} = \frac{20}{49}.$$

4-G **Solution:**

(a) When 6 are selected at random, Y is hypergeometric: The population size is $N = 12$, the number of men in the population is $M = 6$, and the sample size is $n = 6$. Then

$$P(Y = 3) = \frac{\binom{6}{3}\binom{6}{3}}{\binom{12}{6}} = .4329.$$

(b) Let X denote the number assigned to the treatment group; assuming an ideal coin ($p = .5$), this number is binomial: $X \sim \text{Bin}(12, .5)$. Then

$$P(X = k) = \binom{12}{k}\left(\frac{1}{2}\right)^k\left(\frac{1}{2}\right)^{12-k},$$

and with $k = 6$, this probability is $924/4096 \doteq .2256$.

(c) Think of this as a succession of two binomial experiments—one, in which women are assigned (to one group or the other), and the other in which men are assigned. The sequence of coin tosses is a Bernoulli process, and the two experiments are independent. Hence,

$P(3 \text{ men and } 3 \text{ women get the drug})$

$= P(3 \text{ men get the drug}) \times P(3 \text{ women get the drug}) = \left\{\binom{6}{3}(.5)^6\right\}^2 \doteq .098.$

4-H **Solution:**

Think of this result as first getting exactly one black one in the first 4 tries, and then getting a black one in the 5th try:

$$P(Z = 5) = \frac{\binom{8}{1}\binom{42}{3}}{\binom{50}{4}} \times \frac{7}{46} = .06068455,$$

where the first factor is a hypergeometric probability ($N = 50$, $M = 8$, $n = 4$), and the second factor is the *conditional* probability of a black one, given that 7 black ones remain after the first 4 selections.

4-I   **Solution:**
(a) The number of tries is negative binomial—the sum of 2 independent geometric variables with $p = .16$. We can either substitute in the formula for the negative binomial p.f., or just as quickly, reason as in the preceding problem: In the first 4 tries there is exactly one black, followed by a black the 5th try—but now the probability of one black in 4 tries is binomial:

$$P(W = 5) = \binom{4}{1}(.16)^1(.84)^3 \times .16 = .060693.$$

(To four decimal places, this is the same as the answer to 4-H—which is why we kept enough digits to show that the answers are not exactly the same.)

(b) The expected number to get the first black jelly bean is $1/p = 1/.16$; to get two, it takes twice as many, on average: $2/.16 = 12.5$.

4-J   **Solution:**
Your hand calculator with a "$_nC_r$" key may work, to give .27454 as the numerical value. Otherwise, you can write out each factor and do the multiplication. For instance, the denominator is the ratio of 8 factors starting with 1000 (going down in steps of 1) divided by 8!. The binomial approximation is based on the fact that when $N = 1000$ is much larger then $n = 8$, we can get an approximate value as the probability of 4 successes in 8 trials, with $p = 500/1000$:

$$P(4 \text{ successes}) = \binom{8}{4}(.5)^8 = .27344.$$

4-K   **Solution:**
(a) The number required is Negbin(4, .45), with mean $r/p = 4/.45$, or about 8.9.

(b) We could find the probability that it takes k trials, a negative binomial probability for each k from 4 to 12. This is pretty tedious. A simpler approach is to rephrase the question in terms of the number of successes in 12 trials, which is binomial: It will require at most 12 trials if, in the first 12 trials there are at least 4 successes; this can be read in Table IVb, in the block for $n = 12$ at $c = 4$: .8655.

4-L   **Solution:**
Roll once; whatever number comes up, it takes (on average) $1 \div (5/6)$ rolls to get something different; now having seen two different numbers, it takes $1 \div (4/6)$ rolls to get a number different from the first two; etc. The total number of rolls, on average, is thus

$$1 + \frac{1}{5/6} + \frac{1}{4/6} + \frac{1}{3/6} + \frac{1}{2/6} + \frac{1}{1/6} = 14.7.$$

4-M **Solution:**

The selection of viewers is ordinarily done without replacement, in which case the number in the sample who are watching has a hypergeometric distribution. But a "metropolitan area" is apt to have millions of viewers, and we assume that this is the case. Then, because 1,000 is such a small fraction of the population, we can get good approximations to probabilities if we use the binomial approximation, with p defined as the population proportion who are watching the program, or .35:

$$P(\text{at most } 330) = \sum_{k=0}^{330} \binom{1000}{k}(.35)^k(.65)^{100-k}.$$

It is impractical to calculate this sum of 331 terms, and we appeal to a further approximation—the approximation of binomail probabilities by normal curve areas. For this, we need the mean and standard deviation of the number of "successes": $\mu = np = 350$, and $\sigma^2 = npq = 227.5$. Since $npq > 5$, we use (1) of §4.5 to approximate the desired probability:

$$P(\text{at most } 330) \doteq \Phi\left(\frac{330.5 - 350}{\sqrt{227.5}}\right) = \Phi(-1.293) = .0980.$$

4-N **Solution:**

We are being asked to find $P(Y = 32)$ when $Y \sim \text{Bin}(60, .5)$. We can find this probability using a hand calculator:

$$P(Y = 32) = \binom{60}{32}(.5)^{32}(.5)^{28} = \frac{60!}{32!28!} \cdot (5)^{60} = .08996.$$

But here is an opportunity to check the accuracy of the approximation that uses the normal curve. To do this, we need the mean and standard deviation of Y: $np = 30$, and $npq = 15$, so $\sigma_Y = \sqrt{15} = 3.873$.

The normal table is cumulative, and the approximation (1) is given for events of the type $Y \leq 32$. To use it for a single value of Y, we express such a value k as a set difference: $[Y \leq 32] - [Y \leq 31]$, and obtain

$$P(Y = 32) \doteq \Phi\left(\frac{32.5 - 30}{3.873}\right) - \Phi\left(\frac{31.5 - 30}{3.873}\right) = .7407 - .6507 = .0900.$$

Because we are taking a difference, omitting the continuity correction doesn't do too badly, either:

$$\Phi\left(\frac{32 - 30}{3.873}\right) - \Phi\left(\frac{31 - 30}{3.873}\right) = \Phi(.5164) - \Phi(.258) = .6972 - .6019 = .0953.$$

4-O **Solution:**

We can find this (assuming the population size is much larger than the sample size of 20) as a binomial probability:

$$P(Y \geq 18) = \binom{20}{18}.9^{18}.1^2 + \binom{20}{19}.9^{19}.1 + .9^{20} = .677.$$

But if we want to approximate it, the normal curve doesn't do so well ($npq = 20 \times .9 \times .1 = 1.8$). With $p = P(\text{nonwhite}) = .1$ (fairly small) and $n = 20$ (fairly large), a Poisson approximation with $np = 2$ works better:

$$P(Y \geq 18) = P(20 - Y \leq 2) \doteq e^{-2}(1 + 2 + 2^2/2!) = .6767.$$

This is given in Table IV ($m = np = 2$, $c = 2$) as .677. [Compare with the normal approximation, with $\sigma_Y = \sqrt{npq} = \sqrt{1.8}$:

$$P(20 - Y \leq 2) \doteq \Phi\left(\frac{2.5 - 2}{1.34}\right) = .645.]$$

4-P   Solution:

(a) $\mu = np = 200 \times .3 = 60$.

(b) Since $npq = 42$, use a normal approximation: $\Phi\left(\dfrac{49.5 - 60}{\sqrt{42}}\right) = .0526$.

(c) Here, $npq = 3.92 < 5$; but n is large and p is small—use Poi(4):
$P(\text{none}) = e^{-4} = .018$.

(d) Again using Poi(4): $P(\text{at most } 3) = e^{-4}(1 + 4 + 4^2/2 + 4^3/3!) = .4335$.
[Table IV ($m = 4$, $c = 3$) gives .433.]

When p is small (or q small), we recommend using Poisson; if m is outside the range of the table, then go to the normal approximation.

4-Q   Solution:

(a) Let X denote the number of flaws in a 10-square-yard piece. The average number in a piece of this size is 1/5 of the average in a 50-square-yard piece, or $m = .2$:

$$P(X = 0 \text{ or } 1) = f(0) + f(1) = e^{-.2}(1 + .2) = .982.$$

(This is available in Table IV: $m = .2$, $c = 1$.)

(b) Using the definition of conditional probability, and the fact that the event $[X = 1]$ is a subset of $[X \leq 1]$, we have

$$P(X = 1 \mid X \leq 1) = \frac{P(X = 1 \text{ and } X \leq 1)}{P(X \leq 1)} = \frac{P(X = 1)}{P(X \leq 1)} = \frac{.164}{.982} = .166.$$

4-R   Solution:

(a) The expected number in 30 min. is $24 \times 1/2 = 12$. Table IV gives the probability of at most 10 as .347 ($m = 12$, $c = 10$).

(b) The mean in 1 minute is $24 \times 1/60 = .4$; with $m = .4$, $c = 1$, we find $P(0 \text{ or } 1) = e^{-.4}(1 + .4) = .938$ (with a calculator or in Table IV).

(c) The time will exceed 1 min. if and only if there is at most one arrival in the 1-minute period, so this is the same event as in (b): .938.

4-S   Solution:

This is of the form $\sum g(x)f(x) = E[g(X)]$, where f is the p.f. of Poi(4), and $g(x) = x^2$. So,

$$\sum_0^\infty x^2 \cdot \frac{4^x}{x!} e^{-4} = E(X^2) = \text{var } X + (EX)^2 = 4 + 4^2 = 20.$$

4-T   Solution:

Using the definition of conditional probability, we obtain the conditional p.f. as

$$P(X = x \mid X + Y = c) = \frac{P(X = x,\ X + Y = c)}{P(X + Y + c)} = \frac{P(X = x,\ Y = c - x)}{P(X + Y = c)}.$$

As last written, the numerator is the probability of the intersection of independent events, which factors into $P(X = x) \times P(Y = c - x)$. And the three probabilities in the last fraction are then Poisson probabilities, Poi(λ) in the numerator, and Poi(2λ) in the denominator:

$$P(X = x \mid X + Y = c) = \frac{\left(\frac{\lambda^x}{x!} e^{-\lambda}\right)\left(\frac{\lambda^{c-x}}{(c-x)!} e^{-\lambda}\right)}{\frac{(2\lambda)^c}{c!} e^{-2\lambda}} = \binom{c}{x}\left(\frac{1}{2}\right)^x \left(\frac{1}{2}\right)^{c-x}.$$

The final result is a binomial p.f. ($n = c$, $p = 1/2$). Given that there are c arrivals in the two minutes, the number in the first minute is Bin(c, .5); that is, the arrivals are distributed between the two 1-minute periods as though we had tossed a coin to see in which period an arrival falls.

4-U   Solution:

The joint distribution of (B, O, Y, G, T) is multinomial.

(a) $P(4 \text{ of each}) = \binom{20}{4,\ 4,\ 4,\ 4,\ 4}.4^4.2^4.2^4.1^4.1^4 \doteq .00020.$

(b) There are 18 not green or tan; for any i, j, k such that $i + j + k = 18$,

$$P(B = i, O = j, Y = k \mid G = 0, T = 2) =$$

$$= \frac{P(B = i, O = j, Y = k, G = 0, T = 2)}{P(G = 0, T = 2)} = \frac{\frac{20!}{i!j!k!0!2!}.4^i.2^j.2^k.1^0.1^2}{\frac{20!}{0!2!18!}.1^0.1^2.8^{18}}$$

$$= \frac{18!}{i!j!k!}.5^i.25^j.25^k.$$

This is the p.f. of a trinomial distribution with category probabilities in proportion 4:2:2. This could have been predicted: Knowing that you have no green's and 2 tan's, there are 18 that are to be distributed among the other three colors—in the original proportions of those colors.

(c) Since 15 of the 20 are accounted for, only 5 are unknown—in the colors green and tan. With green and tan in equal proportions, the number of green is binomial with $n = 5$ and $p = .5$. [You could go through the kind of formal calculation shown in (b) to draw the same conclusion.]

4-V   **Solution:**

(a) The marginal distribution of X is hypergeometric:

$$P(X = 4) = \frac{\binom{225}{4}\binom{275}{4}}{\binom{500}{8}} = .2647.$$

(b) We use the Bin(8, .4) to approximate the probability in (a):

$$P(X = 4) \doteq \binom{8}{4}(.45)^4(.35)^4 = .2627.$$

(c) The sample space is one of combinations of 8 chosen from 500, and this count gives us the denominator; for the numerator, we count the number of ways of getting 4 O's, 2 A's, 1 B, and 1 AB:

$$P(X = 4, Y = 2, Z = 1) = \frac{\binom{225}{4}\binom{200}{2}\binom{50}{1}\binom{25}{1}}{\binom{500}{8}} = .0282.$$

(d) For the multinomial approximation, we use .45, .40, .10, and .05 as probabilities of types O, A, B, AB, respectively:

$$\frac{8!}{4!2!1!1!}(.45)^4(.4)^2(.1)^1(.05)^1 = .0276.$$

(e) Since $X + Y + Z$ is the number that are not type AB, the probability that it does not exceed 6 is the same as the probability that the number of type AB's (8 minus the number that are not AB) is at least 2:

$$\sum_{2}^{8} \binom{8}{k}(.05)^k(.95)^{8-k}.$$

This is given in the binomial table (IIb) as .0572 [$n = 8$, $p = .05$, $c = 2$]. But we can (since asked to do so) use the Poisson approximation, with parameter $m = np = .4$; Table IV gives $1 - .938 = .062$—right "ball park" but not very accurate, since n is not very large.

4-W   **Solution:**

(a) Differentiating twice, we get

$$\eta'(t) = .4 + .6t + .6t^2 + .4t^3,$$

$$\eta''(t) = .6 + 1.2t + 1.2t^2.$$

Now substitute $t = 1$: $EX = \eta'(1) = 2$, $E[X(X-1)] = 3$. Then

$$\text{var } X = E[X^2 - X] + EX - (EX)^2 = 3 + 2 - 2^2 = 1.$$

(b) The possible values of X are the exponents: 1, 2, 3, 4, and their probabilities are the corresponding coefficients: .4, .3, .2, .1.

4-X **Solution:**

The p.g.f. is

$$\eta(t) = t^0 \cdot \tfrac{9}{13} + (t^1 + t^2 + t^3 + t^4)\tfrac{1}{13}.$$

Differentiating twice, we find

$$\eta'(t) = (1 + 2t + 3t^2 + 4t^3)\tfrac{1}{13}, \quad \eta''(t) = (2 + 6t + 12t^2)\tfrac{1}{13}.$$

Now substitute $t = 1$: $EX = \eta'(1) = 10/13$, and $\eta''(1) = 20/13$. Then

$$\text{var } X = E(X^2) - (EX)^2 = E[X(X-1)] + EX - (EX)^2 = \tfrac{20}{13} + \tfrac{10}{13} - \left(\tfrac{10}{13}\right)^2,$$

or 290/169 (which agrees with what we found earlier, of course).

CHAPTER 5: Additional Problems

Section 5.1

5-A Given that Y has the c.d.f. $F(y) = y/2$ for $0 < y < 2$, find
(a) the value of $F(y)$ when $y > 2$.
(b) $P(.5 < Y < 1.5)$.
(c) $P(Y < .5 \mid Y < 1)$.

5-B Verify that the function $F(x) = \frac{1}{\pi}\left(\frac{\pi}{2} + \text{Arctan } x\right)$ has the properties that are required of a c.d.f. and can serve to define a model for a continuous random variable.

5-C Given that the c.d.f. of X is $\sin x$ for $0 < x < \pi/2$,
(a) find the c.d.f. of the random variable $Y = \sin X$.
(b) find the c.d.f. of the random variable $Z = \sin^2 X$.
(c) find $P(1 < X < 2)$.

5-D Suppose X has the c.d.f. defined as $F(x) = \frac{1}{4}x$ for $0 \le x < 1$, and $F(x) = \frac{1}{4}(x+2)$ for $1 \le x \le 2$. Find the following:
(a) $P(X = 1)$.   (c) $P(X > 0)$.   (e) $P(0 < X < 1)$.
(b) $P(X < 1)$.   (d) $P(-1 < X < .5)$.   (f) $P(0 \le X \le 1)$.

Section 5.2

5-E Suppose X had the p.d.f. $f(x) = \begin{cases} 1/3, \ 0 < x < 1, \\ 2/3, \ 1 < x < 2. \end{cases}$

(a) Find $P(X > 1/2)$.
(b) Find $P(|X - 1| < 1/2)$.
(c) Find and sketch the c.d.f.

5-F Let X have the p.d.f. $f(x) = \frac{1}{2}e^{-|x|}$. Find the following:
(a) $P(|X| > 1)$.   (c) $F_X(x)$.
(b) $P(X > 1 \mid |X| > 1)$.   (d) $f_Y(y)$, where $Y = |X|$.

5-G Given the c.d.f. $F_X(x) = x^{2/3}$, $0 < x < 1$, find
(a) $P(X > .125)$.   (c) the p.d.f. of $Y = X^{1/3}$.
(b) the p.d.f., $f_X(x)$.   (d) the p.d.f. of $V = -\frac{1}{3}\log_e X$.

5-H Let X have the p.d.f. $f(x) = \sqrt{2/\pi}\, x^2 e^{-x^2/2}$, $x > 0$. Use the differential method to approximate this probability: $P(\sqrt{2} < X < \sqrt{2} + .1)$.

Section 5.3-5.5

5-I Given the c.d.f. $F(x) = x^4$, $0 < x < 1$, find
(a) EX.   (b) the median of X.   (c) the 25th percentile.

5-J Given the p.d.f. $f(x) = 6x(1-x)$, $0 < x < 1$, find
(a) EX.   (b) $F(x)$.   (c) the median.   (d) $P(x_{.10} < X < x_{.30})$.

5-K Suppose X has the distribution given in Problem 5-F above. Find
(a) the median and quartiles.   (b) EX.   (c) $E(X^2)$.   (d) $E|X|$.

5-L Given: the random variable Θ has the p.d.f. $f(\theta) = \sin\theta$, $0 < \theta < \pi/2$.
(a) Find the p.d.f. of $U = \cos\Theta$.
(b) Find $E(\cos\Theta)$, two ways—using f_U and using f_Θ.
(c) Find $E(\sin\Theta)$.

Section 5.6

5-M Suppose X is distributed as in Problem 5-J: $f(x) = 6x(1-x)$, $0 < x < 1$.
Find the following:
(a) σ_X.   (b) m.a.d.(X).   (c) var$(1-X)$.   (d) the Z-score for $x = \frac{3}{4}$.

5-N Let $Y = 2(X - \frac{1}{2})$, where X has the distribution in the preceding problem.
(a) Find the mean and standard deviation of Y.
(b) Find $E[(X-1)^2]$.

Sections 5.7-5.8

5-O Let (X, Y) be uniformly distributed over the disc $x^2 + y^2 \le 1$. Give
(a) the joint p.d.f. of the distribution.
(b) the marginal p.d.f.'s.   (d) the distribution of $U = X^2 + Y^2$.
(c) $E(XY)$.   (e) $E(X^2)$.

5-P Consider the joint p.d.f. $f(x, y) = 24xy$, for (x, y) in the triangle bounded
by the coordinate axes and the line $x + y = 1$.
(a) Find the marginal p.d.f.'s, means, and variances.
(b) Find $P(X + Y < 1/2)$.
(c) Calculate $E(XY)$.

5-Q Suppose (X, Y, Z) is uniformly distributed in the unit cube, the product
set of intervals $(0, 1)$ on the three axes.
(a) Give the value of the joint p.d.f. on this support.
(b) Find the marginal distribution of (X, Y).
(c) Find $P(X^2 + Y^2 + Z^2 < 1)$.
(d) Find $E(X^2 + Y^2 + Z^2)$.

Sections 5.9–5.10

5-R Given $EX = 1$, $EY = 2$, $\text{cov}(X, Y) = -1$, var $X = 4$, and var $Y = 1$, find
(a) $E(2X - Y + 4)$.   (c) $\text{cov}(X + Y, 2X - Y)$
(b) $\text{var}(2X - Y + 4)$.   (d) $E[(X + Y)^2]$.

5-S Given the joint distribution of the Problem 5-P, define new variables
$U = X - 2Y$, $V = 3X + Y - 4$.
(a) Find var U and var V.
(b) Find $\text{cov}(U, V)$.
(c) Calculate $\rho_{U, V}$.
(d) Are X and Y independent? (Explain.)

5-T Suppose $X \sim \mathcal{U}(0, 1)$, and the p.d.f. of Y is $f_Y(y) = 2y$, $0 < y < 1$. Find
the joint p.d.f. of (X, Y), given that X and Y are independent.

5-U Show that if X and Y are independent, $E[g(X)h(Y)] = E[g(X)]E[h(Y)]$.

5-V Suppose X and Y are independent, each with support on $(0, 1)$, and let
$U = X$, $V = X - Y$. Can the variables U and V be independent?

5-W Find the p.d.f. of $X + Y$, where X and Y are independent, each $\mathcal{U}(0, 1)$.

Section 5.11

5-X Problem 5-P dealt with the distribution defined by the joint p.d.f.
$f(x, y) = 24xy$, for (x, y) in the triangle bounded by the coordinate axes
and the line $x + y = 1$. Find the conditional density functions of Y given
$X = x$, and of X given $Y = y$.

5-Y Suppose $X \mid Y = y$ is uniform on $(0, y)$, and that Y has the p.d.f. $2y$, for
$0 < y < 1$. Find
(a) the joint p.d.f. of the pair (X, Y).
(b) the marginal p.d.f. of X.
(c) the conditional p.d.f. of Y given $X = x$.
(d) the conditional means.
(e) verify the iterated expectation formula [(6) of §5.11].

5-Z Check the relations (5)-(7) in §5.11 in the special case in which X and Y
are independent.

Section 5.12

5-AA Find the m.g.f. of a random variable X with p.d.f. $\frac{1}{2}e^{-|x|}$. (Referring to Problem 5-69, deduce the p.d.f. of the X in that problem).

5-BB Find the first few moments of the distribution with m.g.f. $(1 - t^2)^{-1/2}$.

5-CC Given that X has the m.g.f. $\psi(t) = \frac{1}{2}(e^t + e^{-t})$, find the p.f. of X.

5-DD Let X have the triangular p.d.f. $f(x) = 1 - |x|$ for $|x| < 1$.
(a) Find the m.g.f. of X.   (b) Use the m.g.f. to find var X.

Chapter 5: Solutions

5-A   Solution:

(a) $F(2) = 1$, and because F cannot decrease nor exceed 1, $F(y) = 1$ for any y greater than 2.

(b) The probability of an interval is the increase in F over that interval:

$$F(1.5) - F(.5) = .75 - .25 = .5.$$

(c) Use the defining formula for conditional probability:

$$P(Y < .5 \mid Y > 1) = \frac{P(Y < .5 \text{ and } Y < 1)}{P(Y < 1)} = \frac{P(Y < .5)}{P(Y < 1)} = \frac{F(.5)}{F(1)} = \frac{1}{2},$$

5-B   Solution:

We know that F is continuous and differentiable everywhere. It is non-decreasing if its derivative is positive, which it is: $F'(x) \propto (1 + x^2)^{-1}$. Moreover, $\text{Arctan}(-\infty) = -\pi/2$, and $\text{Arctan}(\infty) = \pi/2$, so $F(-\infty) = 0$, and $F(\infty) = 1$.

5-C   Solution:

(a) To find a c.d.f. we can always turn to its definition:

$$P(\sin X \le y) = P(X \le \text{Arcsin } y) = F_X(\text{Arcsin } y) = \sin(\text{Arcsin } y) = y,$$

for any y between 0 and 1. That is, $\sin X$ is uniform on (0, 1).

(b) Using the same approach, we have

$$P(\sin^2 X \le z) = P(\sin X \le \sqrt{z}) = P(Y \le \sqrt{z}) = \sqrt{z}, \quad 0 < z < 1.$$

It should be noted that for the intervals involved, the functions $\sin x$ and $\sin^2 x$ are strictly monotonically increasing, with unique inverses, and this is what allows us to take the inverse function of each side of the inequality with the inequality preserved.

(c) As with any interval, $P(1 < X < 2) = F(2) - F(1)$. The function F was only given for the interval $(0, \pi/2)$, but $F(\pi/2) = 1$, so F must be 1 for any x larger than $\pi/2$, such as $x = 2$: $F(2) = 1$. Then, since $F(1) = \sin 1 \doteq .84$, the desired probability is $1 - .84$ or .16. [Note: x in radians.]

5-D   Solution:

This is much simpler when you first draw a graph of F. Do this before proceeding. You'll see that F has a jump of 1/4 at 0 and another jump of 1/4 at 1, and is linear between (0, .25) and (1, .75).

(a) The probability *at* a point is the jump there: 1/4.

(b) Since $F(1) = 1$, $P(X < 1) = P(X \le 1) - P(X = 1) = 1 - 1/4 = 3/4$.

(c) $P(X > 0) = 1 - P(X \leq 0) = 1 - F(0) = 1 - 1/4 = 3/4.$

(d) There is no jump at .5, so this is the same as $P(-1 < X \leq .5)$, or $F(.5) - F(-1) = .5 - 0.$

(e) $P(0 < X < 1) = P(0 < X \leq 1) - P(X = 1) = F(1) - F(0) -$ jump at 1, or $1 - .25 - .25 = .5.$

(f) To the interval in (e), we adjoin the points 0 and 1; this adds in the jump at 0 and the jump at 1: $.5 + .25 + .25 = 1.$

5-E   Solution:

As above, it's a good idea to draw a graph when you can. We've shown it in Figure 4, below.

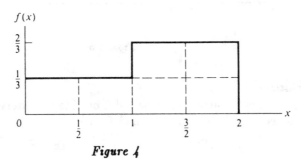

Figure 4

(a) The relative area to the right of $x = 1/2$ is 5/6 of the total; this is the probability called for.

(b) The inequality can be written $-1/2 < X - 1 < 1/2$, or $1/2 < X < 3/2$. So what's wanted is the area between 1/2 and 3/2; this is 1/2 the total, and this is the probability.

(c) We've sketched this in Figure 5. Observe that the area under f between 0 and 2 is 1, so there is no probability outside this interval. This means that the c.d.f. is 0 until you hit $x = 0$ (moving from left to right). Then it rises with constant slope 1/3 (linearly) to 1/3 at $x = 1$; from there, it rises with constant slope 2/3 (linearly) until it reaches 1 at $x = 2$. And that's as high as it gets—$F = 1$ from there on.

We could also find the c.d.f. using a formal integration of the p.d.f., but we hope you would do it in terms of area (after all, integrating is finding area!), when the areas are easy to calculate from geometry. To illustrate the integration process, let's find F for an x on the interval $(1, 2)$:

$$F(x) = \int_{-\infty}^{x} f(u)\,du = \int_{0}^{1} \frac{1}{3}\,du + \int_{1}^{x} \frac{2}{3}\,du = \frac{1}{3} + \frac{2}{3}(x - 1),$$

and this is the equation of the line we've drawn between $x = 1$ and $x = 2$.

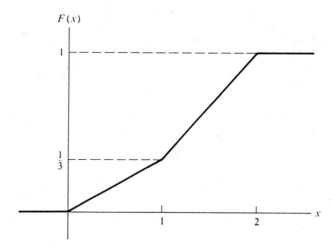

Figure 5

5-F   Solution:
(a) This is the area under the p.d.f. outside the interval $(-1, 1)$, or (by symmetry) twice the area to the left of -1:

$$2P(X < -1) = 2 \int_{-\infty}^{-1} \tfrac{1}{2} e^x \, dx = e^{-1}.$$

(When x is negative, $-|x|$ is just x.)

(b) This you can do without integrating, exploiting symmetry: The area above $X > 1$ is half the area above $|X| > 1$, so the probability is $1/2$ that you end up in $X > 1$, given that you are either in $X > 1$ or $X < -1$.

(c) Because the p.d.f. is given piece-wise, it is necessary to integrate piece-wise. For negative x,

$$F(x) = \int_{-\infty}^{x} \tfrac{1}{2} e^u \, du = \tfrac{1}{2} e^x.$$

For positive x,

$$F(x) = \tfrac{1}{2} + \int_{0}^{x} \tfrac{1}{2} e^{-u} \, du = \tfrac{1}{2} + \tfrac{1}{2}(1 - e^{-x}) = 1 - \tfrac{1}{2} e^{-x}.$$

(d) We'd prefer that you think of this heuristically: Taking the absolute value reflects what is to the left of 0 to the right half-axis; and because of the symmetry, this just doubles what was there originally. So the new p.d.f. is $f_Y(y) = e^{-y}$, $y > 0$.

5-G **Solution:**

(a) Since you already have the c.d.f., which is an integral of the p.d.f., you don't have to integrate the p.d.f. again—just substitute in the c.d.f.:

$$P(X > .125) = 1 - F_X(.125) = 3/4.$$

(b) $f_X(x) = F'_X(x) = \frac{2}{3}x^{-1/3}, \ 0 < x < 1.$

(c) Use the transformation formula (16), and the inverse $x = y^3$:

$$f_Y(y) = f_X(y^3) \cdot 3y^2 = \frac{2}{3}(y^3)^{-1/3} \cdot 3y^2 = 2y, \ 0 < y < 1.$$

(d) The inverse transformation here is $x = e^{-3v}$ (which is 1-1):

$$f_V(v) = f_X(e^{-2v}) \cdot 2e^{-2v} = \frac{2}{3}(e^{-3v})^{-1/3} \cdot 3e^{-3v} = 2e^{-2v}, \ v > 0.$$

(When $0 < x < 1$, $-\frac{1}{3}\log_e x > 0$.)

5-H **Solution:**

We apply (5) of §5.2, with $x = \sqrt{2}$ and $dx = .1$:

$$P(\sqrt{2} < X < \sqrt{2} + .1) \doteq f(\sqrt{2}) \times .1 = .2e^{-1}\sqrt{2/\pi} = .0587.$$

5-I **Solution:**

(a) The p.d.f. is $F'(x) = 4x^3$, so $EX = \int_0^1 x \cdot 4x^3 \, dx = 4/5.$

(b) The c.d.f. is given—set it equal to .5 and solve: $x = .5^{1/4} \doteq .841.$

(c) Set $F(x)$ equal to .25: $x^4 = .25$, $x = (.25)^{1/4} = 1/\sqrt{2} = .707.$

5-J **Solution:**

(a) You can always integrate $xf(x)$: $EX = \int_0^1 6x^2(1-x)\,dx = 1/2,$

but in this case it would be smarter to observe that the p.d.f. is symmetric about 1/2, which is both the mean and the median.

(b) $F(x) = \int_0^x 6u(1-u)\,du = 3x^2 - 2x^3, \ 0 < x < 1.$

(c) We could set $F(x) = .5$, but we already saw in (a) that the median is 1/2, the center of symmetry.

(d) For **any** (continuous) distribution, the probability between two percentiles is the difference in percentages: $.30 - .10 = .20.$

5-K Solution:

(a) Symmetry about 0 means that this is the median. The first quartile is the value of x with (from 5-F) $F(x) = \frac{1}{2}e^x = .25$, or $\log .5 = -.693$. The third quartile is symmetrically located: $+.693$.

(b) The integral for the mean converges, so the mean is 0, the point of symmetry, 0.

(c) Both x^2 and $f(x)$ are even functions (symmetric about 0), so the integral over $(-\infty, \infty)$ is twice the integral over $(0, \infty)$:

$$E(X^2) = 2 \int_0^\infty x^2 \left(\tfrac{1}{2}e^{-x}\right) dx = 2.$$

(d) For the same reason as in (c), $E|X| = 2 \int_0^\infty x \left(\tfrac{1}{2}e^{-x}\right) dx = 1.$

5-L Solution:

(a) We could use the formula for transforming p.d.f.'s, but to avoid differentiating inverse trig functions, let's find $F(\theta)$:

$$F(\theta) = \int_0^\theta \sin x \, dx = 1 - \cos \theta.$$

Then $P(\cos \Theta \leq u) = P(\Theta \geq \cos^{-1} u) = \cos(\cos^{-1} u) = u, \; 0 < u < 1.$
So $f_U(u) = 1, \; 0 < u < 1.$

(b) EU is the mean of $\mathcal{U}(0, 1)$, or $1/2$ (the midpoint of the support).
Or, $EU = \int_0^{\pi/2} \cos\theta \, \sin\theta \, d\theta = \frac{1}{2}\int_0^{\pi/2}\sin 2\theta \, d\theta = 1/2.$

(c) Here it's easier to use the second way:

$$EV = E(\sin \Theta) = \int_0^{\pi/2} \sin\theta \, \sin\theta \, d\theta = \frac{1}{2}\int_0^{\pi/2}(1 - \cos 2\theta) \, d\theta = \pi/4.$$

5-M Solution:

(a) First we find the mean square:

$$E(X^2) = \int_0^1 x^2 \cdot 6x(1-x) \, dx = \frac{6}{20}.$$

The mean is $1/2$, so the $\sigma^2 = 3/10 - 1/4 = 1/20$, and $\sigma = 1/\sqrt{20} \doteq .224.$

(b) The mean deviation (about the mean) is

$$E|X - .5| = \int_0^1 |x - .5| \cdot 6x(1-x) \, dx.$$

Both $f(x)$ and the absolute deviation $|x - .5|$ are symmetric about .5: the graph of f is a parabola (quadratic function), opening down, whose maximum is at .5. This means that we can find the m.a.d. as twice the integral from 0 to .5, where $|x - .5|$ is $.5 - x$:

$$\text{m.a.d.} = 2\int_0^{.5}(.5 - x)\cdot 6x(1 - x)\,dx = .5 - 2\int_0^{.5}6x^2(1 - x)\,dx = \tfrac{3}{16}.$$

(c) $\text{var}(1 - X) = \text{var}\,X = 1/20$ [from (a)].

(d) $Z = \dfrac{X - \mu}{\sigma} = \dfrac{3/4 - 1/2}{1/\sqrt{20}} \doteq 1.12$, so 3/4 is about 1.12 s.d.'s to the right of the mean.

5-N Solution:

(a) The s.d. of $X - 1$ is the same as the s.d. of $2X$, which is $2\sigma_X = .447$, from 5-L(a). The mean is $2(EX - \tfrac{1}{2}) = 2EX - 1 = 0$.

(b) From the parallel axis theorem: $\text{var}\,X + (EX - 1)^2 = \tfrac{1}{20} + \tfrac{1}{4} = \tfrac{3}{10}$.

5-O Solution:

(a) The p.d.f. is a constant; the value of the constant is the reciprocal of the area of the support: $1/\pi$ (the volume under the p.d.f. must be 1).

(b) To find f_X, we integrate out the y across the chord of the circle at x, that is, between the limits $a = \pm\sqrt{1 - x^2}$:

$$f_X(x) = \int_{-a}^a \tfrac{1}{\pi}\,dy = 2\sqrt{1 - x^2}.$$

The p.d.f. is a symmetric function of x and y, so the variables are exchangeable, which means that the p.d.f.'s are the same: $f_Y(y) = f_X(y)$.

(c) Don't integrate—just observe the symmetry. The products $xy\,dx\,dy$ in one quadrant just cancel those in an adjacent quadrant: $E(XY) = 0$.

(d) If you don't see how to approach finding a distribution, try finding the c.d.f.:

$$F_U(u) = P(X^2 + Y^2 < u).$$

This probability is the volume under the joint p.d.f. above a circle of radius $\sqrt{u}$: height × area of base $= \tfrac{1}{\pi}\times\pi(\sqrt{u})^2 = u$, for $0 < u < 1$. So $U \sim \mathcal{U}(0, 1)$.

(e) From (d), we know that $E(X^2 + Y^2) = E(X^2) + E(Y^2) = 1/2$, and because X and Y are exchangeable, $E(X^2) = 1/4$.

5-P   **Solution:**

(a)   To find f_X, we integrate out the y, holding x fixed, as y varies across the triangular support from $y = 0$ to the line where $y = 1 - x$:

$$f_X(x) = \int_0^{1-x} 24xy \, dy = 12x \int_0^{1-x} 2y \, dy = 12x(1-x)^2, \ 0 < x < 1.$$

The mean and mean square are

$$EX = \int_0^1 x \cdot 12x(1-x)^2 dx = 2/5, \quad E(X^2) = \int_0^1 x^2 \cdot 12x(1-x)^2 dx = 1/5.$$

But X and Y are exchangeable, so $EY = EX$, and var $Y =$ var $X = 1/25$.

(b)   This is the double integral of the joint p.d.f. over the region (within the support) where $x + y < 1/2$:

$$\int_0^{1/2} \left\{ \int_0^{1/2 - y} 24xy \, dx \right\} dy = \int_0^{1/2} 12y(\tfrac{1}{2} - y)^2 dx = 1/16.$$

(c) $E(XY) = \int_0^1 8x^2 \left\{ \int_0^{1-x} 3y^2 \, dy \right\} dx = 8 \int_0^1 x^2 (1-x)^3 dx = 2/15.$

5-Q   **Solution:**

(a)   The density is a constant, chosen so that the triple integral over the support region is 1. Because the density is constant, the triple integral is just the volume of the unit cube, which is 1, times the constant—which must also be 1.

(b)   The marginal p.d.f. of (X, Y) is obtained by integrating the p.d.f. over Z from 0 to 1. The integral of 1 over $(0, 1)$ is 1. So the marginal joint distribution of X and Y is uniform on the unit square.

(c)   The probability of a subregion of the supporting cube is the volume of that subregion divided by the volume of the cube, because the density is uniform. The volume of the sphere with radius 1 within the first octant is : $\tfrac{1}{8} \times \tfrac{4}{3}\pi r^3 = \pi/6$, so the desired probability is $\pi/6 \div 1 = \pi/6$.

(d)   The variables are exchangeable, so what's wanted is $3E(X^2)$, where X is uniform on the unit interval. [Imagine the probability as a cloud of dust uniformly distributed throughout the cube, and what distribution results when all the dust is projected onto the x-axis.] We've calculated $E(X^2)$ before (Example 5.6a)—it is 1/3. So $E(X^2 + Y^2 + Z^2) = 1$.

5-R   **Solution:**

(a)   Averaging is a linear operation: $2EX - EY + 4 = 4$.

(b)   The added 4 does not affect the variance, so this is var$(2X - Y)$, or var$(2X) +$ var$(-Y) - 2\,$cov$(2X, -Y) = 16 + 1 - 4(-1) = 21$.

(c) $\text{cov}(X, 2X) + \text{cov}(Y, -Y) + \text{cov}(X, Y) = 8 - 1 - 1 = 6.$

(d) This is $\text{var}(X + Y) + (EX + EY)^2 = 3 + 3^2 = 12.$

5-S   Solution:

(a) From the 5-P, $\text{cov}(X, Y) = E(XY) - EX \cdot EY = \frac{2}{15} - \frac{2}{5} \cdot \frac{2}{5} = -\frac{2}{75}$, so

$\text{var } X + \text{var }(-2Y) + 2\,\text{cov}(X, -2Y) = \sigma_X^2 + 4\sigma_Y^2 - 4\sigma_{X,Y} = \frac{5}{25} + \frac{8}{75} = \frac{23}{75}$,

$\text{var}(3X) + \text{var } Y + 2\,\text{cov}(3X, Y) = 10\sigma_X^2 + 6\sigma_{X,Y} = \frac{10}{25} - \frac{12}{75} = \frac{6}{25}.$

(b) $\text{cov}(X - 2Y, 3X + Y) = 3\sigma_X^2 - 2\sigma_Y^2 - 5\sigma_{X,Y} = \frac{1}{25} + \frac{10}{75} = \frac{13}{75}.$

(c) This is (b) divided by the square root of the product of the variances:

$$\rho_{U,V} = \frac{13/75}{\sqrt{\frac{6}{25} \times \frac{23}{75}}} = \frac{13}{\sqrt{18 \times 23}} \doteq .64.$$

(d) Dependence is evident in the fact that the support is a triangle—not a product set. (It would also follow from the fact that $\sigma_{X,Y} \neq 0$.)

5-T   Solution:
When variables are independent, the joint p.d.f. is the product of the marginal p.d.f.'s:

$$f(x, y) = f_X(x)f_Y(y) = 1 \times 2y = 2y, \text{ for } 0 < x < 1, \, 0 < y < 1.$$

5-U   Solution: Independence means that the joint p.d.f. is the product of the marginal p.d.f.'s: $f(x, y) = f_X(x)f_Y(y)$. Then

$$E[g(X)h(Y)] = \int \int g(x)h(y)\,f_X(x)f_Y(y)\,dx\,dy$$

$$= \int h(y)f_Y(y)\left\{\int g(x)f_X(x)\,dx\right\}dy.$$

The x-integral is $E[g(X)]$—a constant, which factors out of the y-integral, and the y-integral is $E[h(Y)]$.

5-V   Solution:
The boundary lines of the support of (X, Y) are the lines $x = 0$, $x = 1$, $y = 0$, and $y = 1$. Solving for X and Y, we find $X = U$, and $Y = U - V$. The images of the boundary lines give the boundary of the support of U and V: The image of $x = 0$ is $u = 0$, the image of $x = 1$ is $u = 1$, the image of $y = 0$ is $u = v$, and the image of $y = 1$ is $u - v = 1$. So the support of the distribution of (U, V) is within the strip $0 < x < 1$, and bounded by the lines $v = u$ and $v = u - 1$. The latter are $45°$ lines—the support is a trapezoid, one that is not a rectangle; so U and V cannot be independent.

5-W Solution:

We'll use a standard approach: Find the c.d.f. of the sum, and then differentiate to find the p.d.f. Let $Z = X + Y$. Then

$$F_Z(z) = P(Z < z) = P(X + Y < z) = \int\int_A f(x, y)\, dx\, dy,$$

where A is that part of the support (in this instance, the unit square) where $x + y < z$. This region is shown shaded in Figure 6 for the case in which $1 < z < 2$.

Rather than integrate, however, we observe that the joint p.d.f. is 1 on the unit square—the joint distribution is *uniform* there, so probability is proportional to area. The area of the shaded region is easy to find as 1 minus the area of the unshaded triangle: $1 - \frac{1}{2}(2 - z)^2$. For the case $0 < z < 1$, the appropriate region is just a 45° triangle, with legs of length z, and area $\frac{1}{2}z^2$. To find the p.d.f., we differentiate these two functions of z, obtaining

$$f_Z(z) = \begin{cases} z, & 0 < z < 1, \\ 2 - z, & 1 < z < 2. \end{cases}$$

This is a triangular distribution centered at 1, between 0 and 2.

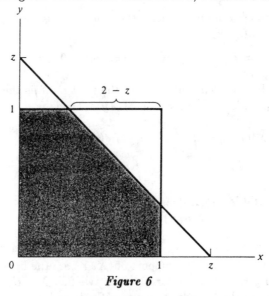

Figure 6

5-X Solution:

The defining formula for $Y \mid x$ (with f_X from 5-P) gives us

$$f(y \mid x) = \frac{f(x, y)}{f_X(x)} = \frac{24xy}{12x(1 - x)^2} = \frac{2y}{(1 - x)^2}, \quad 0 < y < 1 - x.$$

Because the variables are exchangeable, to find $f(x \mid y)$ we just interchange

x and y in the above formula for $f(y \mid x)$.

To check that $f(y \mid x)$ is a p.d.f., we'll find its integral:

$$\int_0^{1-x} \frac{2y}{(1-x)^2} dy = \frac{1}{(1-x)^2} \int_0^{1-x} 2y \, dy = 1.$$

5-Y   Solution:

(a) Since $f(x \mid y) = f(x, y)/f_Y(y)$, it follows that
$$f(x, y) = f(x \mid y) f_Y(y) = \frac{1}{y} \times 2y = 2 \quad \text{for } 0 < x < y < 1.$$

(b) Integrating out the y produces f_X:
$$f_X(x) = \int_x^1 2 \, dy = 2(1-x), \ 0 < x < 1.$$

(c) Dividing (a) by (b) results in the conditional p.d.f. of Y given $X = x$:
$$f(y \mid x) = \frac{2}{2(1-x)} = \frac{1}{1-x}, \quad x < y < 1.$$
This says that Y is uniform on $(x, 1)$. [Observe that we now have another way of seeing that X and Y are not independent: The product of the marginal p.d.f.'s is clearly not equal to the joint p.d.f.]

(d) Both conditional distributions are uniform, so the conditional means are the midpoints of the intervals of support: $E(X \mid y) = y/2$, and $E(Y \mid x) = (1+x)/2$.

(e) $E(E[X \mid Y]) = E[Y/2] = \frac{1}{2} \int_0^1 y(2y) \, dy = \frac{1}{3} = \int_0^1 x \cdot 2(1-x) \, dx = EX.$

Also,
$$E(E[Y \mid X]) = \frac{1}{2} E(1 + X) = \frac{1}{2} + \frac{1}{2} EX = \frac{2}{3} = EY.$$

5-Z   Solution:

In this case, $f(y \mid x) = \dfrac{f(x, y)}{f_X(x)} = f_Y(y)$; then $E[f(y \mid X)] = f_Y(y)$.

The conditional mean of Y given $X = x$ is just the mean of Y, which is a constant; so $E[E(Y \mid X)] = EY$. And, because $E(Y \mid X)$ is constant, its variance is 0, so both sides of (7) are just var Y.

5-AA   Solution:

We use the definition and calculate the expected value as an integral:

$$E(e^{tX}) = \int_{-\infty}^{\infty} e^{tx} \cdot \frac{1}{2} e^{-|x|} \, dx = \frac{1}{2} \int_{-\infty}^0 e^{x + tx} dx + \frac{1}{2} \int_0^{\infty} e^{-x + tx} dx$$

$$= \frac{1}{2} \left\{ \frac{1}{1+t} + \frac{1}{1-t} \right\} = \frac{1}{1-t^2}.$$

(Since this is just the m.g.f. from Problem 5-69, and because only one distribution has a given m.g.f., the distribution of X in that problem must have the p.d.f. assumed in this problem. Also, as a manifestation of the property shown in Problem 5-74, observe that ψ is symmetric about 0.)

5-BB Solution:
We could differentiate a few times, but perhaps it's just as easy (if not easier) to use the negative binomial theorem to expand:

$$[1 + (-t^2)]^{-1/2} = 1^{-1/2} + (-\tfrac{1}{2})1^{-3/2}(-t^2)^1 + \frac{(-\tfrac{1}{2})(-\tfrac{3}{2})}{2!}1^{-5/2}(-t^2)^2 + \cdots$$

$$= 1 + t^2/2 + 3t^4/8 + \cdots.$$

All odd-order moments are 0 because the coefficients of odd powers of t are all 0. The 2nd moment is the coefficient of $t^2/2$, or $E(X^2) = 1$. The 4th moment is the coefficient of $t^4/4!$, or $E(X^4) = 9$.

5-CC Solution:
The definition of ψ is $\psi(t) = \sum_{x_i} e^{tx_i} f(x_i)$. Comparing this with the given m.g.f., we can make the identification with $x_1 = 1$ with probability $f(x_1) = \tfrac{1}{2}$, and $x_2 = -1$ with probability $f(x_2) = \tfrac{1}{2}$. (Because there is only one distribution with a given m.g.f., this is the p.f. we seek.)

5-DD Solution:
(a) We could integrate $e^{tx} f(x)$, but we have essentially found the result in Problem 5-W, if we recall from Example 5-12e that the m.g.f. of $\mathcal{U}(0, 1)$ is $(e^t - 1)/t$. The given triangular distribution is the same as that found in 5-W except that it is centered at 0; the Z of that problem is $X + 1$ of this problem: $X = Z - 1$, where the m.g.f. of Z is $[(e^t - 1)/t]^2$:

$$\psi_X(t) = E e^{t(Z-1)} = e^{-t}\psi_Z(t) = e^{-t} \times \frac{(e^t - 1)^2}{t^2}.$$

(b) Differentiating this twice is a bit of a challenge—better to use series. First write it as

$$\frac{e^t - 2 + e^{-t}}{t^2} = \frac{2(\cosh t - 1)}{t^2}.$$

The series for the hyperbolic cosine is $1 + t^2/2 + t^4/4! + \cdots$, so

$$\psi_X(t) = 1 + t^2/12 + \cdots,$$

and $EX = 0$, and var $X = E(X^2)$, the coefficient of $t^2/2$, or $1/6$.

CHAPTER 6: Additional Problems

Section 6.1

6-A Let X denote the net weight in ounces of a "half-pound" bag of potato chips. Assume that X is approximately normal with mean 8 oz and s.d. 1/4 oz. Find
(a) $P(X < 7.5)$.
(b) $E(X^2)$.
(c) the value of c such that 90% of the bags weigh between $8 \pm c$ oz.

6-B Given that X has the p.d.f. $f(x) \propto \exp[-4x^2 + 6x]$, find
(a) EX.   (b) var X.   (c) the constant of proportionality.

6-C Given that $X \sim N(\mu, \sigma^2)$, find $E(X^4)$.

6-D Given that X is normal, and that $x_{.1} = 15$ and $x_{.6} = 25$, find μ and σ.

6-E Given that X and Y are independent, $X \sim N(10, 3)$ and $Y \sim N(6, 9)$, find the distributions of the following:
(a) $X + Y$.   (b) $X - Y$.   (c) $2X + 5Y$.

6-F Suppose the lengths of 1800-foot audio tapes of a certain brand are normal with mean 1800 ft and s.d. 3 ft. Find the probability that the lengths of 2 rolls of tape differ by more than 8 ft.

Sections 6.2-6.3

6-G Suppose calls arrive at an exchange according to a Poisson law at the rate 8/min. Let T denote the time, measured from an arbitrary point in time, to the next call. Find
(a) the probability that T exceeds 9 seconds.
(b) the mean and s.d. of T.
(c) the mean time between the most recent call and the next call.
(d) the expected waiting time to the 10th call.
(e) the probability that the time to the 10th call exceeds 1 minute.

6-H Suppose 3 different types of customers are served at a facility, each arriving according to a Poisson process: type A at the rate 10/hr, type B at the rate 15/hr, and type C at the rate 5/hr. Assuming independence of the three processes, find
(a) the mean time to the 10th arrival of any customer.
(b) how long, on average, you would have to wait for the 5th arrival of a type A customer.
(c) the p.d.f. of T, the time to the first arrival of either type A or type C, and the mean value of T.
(d) the probability that a customer of type A arrives before one of type B.

6-I Find the numerical value of each of these integrals:

(a) $\displaystyle\int_0^\infty x^{13} e^{-3x}\, dx.$ (b) $\displaystyle\int_0^\infty x^{9/2} e^{-x}\, dx.$

6-J Suppose X has the "Rayleigh distribution," with p.d.f. defined as
$$f(x\,|\,\theta) = \theta x e^{-\theta x^2/2}, \quad x > 0.$$
Write the integral that gives the mean value, and make the change of variable $\theta x^2/2 = u$ to express the integral in terms of a gamma function, using this to find the mean.

Sections 6.4-6.5

6-K Find an approximate value of the 90th percentile of $\text{chi}^2(25)$,
(a) using the fact that $\chi^2 \approx N(k, 2k)$.
(b) using the fact that $\sqrt{2\chi^2} - \sqrt{2k-1} \approx N(0, 1)$.

6-L Find $E(1/X)$ when $X \sim \text{chi}^2(k)$.

6-M Suppose we do an experiment which yields X, assumed to be $N(\mu, \sigma^2)$, ten times, in independent trials, resulting in the random sample $(X_1, ..., X_{10})$.

Find $P\left\{\displaystyle\sum_1^{10}\left(\frac{X_i - \mu}{\sigma}\right)^2 \geq 16\right\}.$

6-N Suppose a certain type of unit has an exponential time to failure, with mean 2 hrs. Find the following:
(a) The probability that one unit lasts at least 4 hrs.
(b) The probability that at least one of a pair of units operating independently lasts 4 hrs, i.e., $P(Y > 4)$, where Y is the longer life time.
(c) The mean value of Y in (b).
(d) $E(W)$, where W is the time to the *first* failure of the pair in (b).

6-O Three units are linked in a system as shown in Figure 7: The system fails if either units 1 and 2 both fail, or units 2 and 3 both fail. Find the reliability function of the system in terms of R_i, the reliability function of unit i.

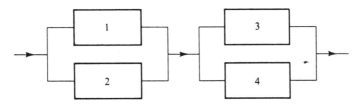

Figure 7

6-P The Weibull distribution with parameters $\alpha = 2$ and $\beta = 1$ [§6.7, p. 250] has the reliability function $R(t) = \exp(-t^2)$. Given that T has this distribution, find

(a) $P(T < 1)$.　　　(b) ET.　　　(c) var T.

Chapter 6: Solutions

6-A   Solution:

(a) The Z-score for $X = 7.5$ is $Z = \dfrac{7.5 - 8}{.25} = -2$. From Table IIa, we see that the area to the left of -2 is .0228, which is the desired probability.

(b) The parallel axis theorem says that $E(X^2) = \sigma^2 + \mu^2 = 64.0625$.

(c) The given interval is symmetric about the mean, so there would have to be 5% of the distribution to the left of $8 - c$ and 5% to the right of $8 + c$. The 95th percentile of Z (from Table IIa) is 1.645, so

$$8 + c = \mu + 1.645\sigma, \text{ and } c = 1.645\sigma = .411.$$

6-B   Solution:

To identify this with the formula for the general normal p.d.f. [(7) of §6.1], we complete the square in the exponent. First, factor out the -4:

$$-4x^2 + 6x = -4(x^2 - 3x/2 \qquad).$$

To complete the square, we fill in the blank space with the square of one-half the coefficient of x, or 9/16. To preserve the balance, we need also to add $4 \times 9/16 = 9/4$:

$$-4x^2 + 6x = -4(x^2 - 3x/2 + 9x^2/16) + 9/4.$$

So the density is

$$f(x) = Ke^{9/4} \cdot e^{-4(x - 3/4)^2}.$$

Now, comparing with (7), we see that $4 = 1/(2\sigma^2)$, $3/4 = \mu$, and $Ke^{9/4}$ is $1/\sqrt{2\pi\sigma^2}$. Thus, $\sigma^2 = 1/8$, and $K = e^{-9/4}/\sqrt{2\pi\sigma^2}$.

6-C   Solution:

We do know the 4th central moment: $E[(X - \mu)^4] = 3\sigma^4$, as well as the 2nd central moment, σ^2. Also, all odd-order central moments are 0 [from (10) of §6.1]. Now write X as $X - \mu + \mu$, and expand:

$$X^4 = [(X - \mu) + \mu]^4 = (X - \mu)^4 + 6(X - \mu)^2\mu^2 + \mu^4$$

$$+ \text{odd powers of } X - \mu.$$

Taking expected values on both sides we obtain

$$E[X^4] = 3\sigma^4 + 6\sigma^2\mu^2 + \mu^4.$$

6-D **Solution:**

First we find the 10th and 60th percentiles of the standard normal Z:
$z_{.1} = -1.2826$ and $z_{.6} = .2533$ (from Table IIa). Then, write the relations
between x and z $(x = \mu + \sigma z)$, for the given percentiles:

$$\begin{cases} 15 = \mu - 1.2816\,\sigma \\ 25 = \mu + .2533\,\sigma \end{cases}$$

Subtracting the first equation from the second, we get $10 = 1.5349\,\sigma$, or
$\sigma = 6.515$; substituting in either equation we find $\mu = 23.35$.

6-E **Solution:**

We know that linear combinations of independent normal variables are
normal. It remains only to find the means and variances, which complete
the characterization of the distributions.

(a) $\mu = 10 + 6 = 16$, $\sigma^2 = 3 + 9 = 12$.

(b) $\mu = 10 - 6 = 4$, $\sigma^2 = 3 + 9 = 12$.

(c) $\mu = 2 \times 10 + 5 \times 6 = 50$, $\sigma^2 = 4 \times 3 + 25 \times 9 = 237$.

6-F **Solution:**

We assume the lengths can be considered independent, in which case the
difference in lengths is normal with mean 0, and variance $3^2 + 3^2 = 18$.
Then,

$$P(|L_1 - L_2| > 8) = 1 - P(-8 < L_1 - L_2 < 8)$$
$$= 1 - \Phi\left(\frac{8-0}{\sqrt{18}}\right) + \Phi\left(\frac{-8-0}{\sqrt{18}}\right) = .0592.$$

6-G **Solution:**

(a) With the rate given in minutes, we convert 9 seconds to 9/60 minutes.
Then, since $T \sim \text{Exp}(8)$, $P(T > .15) = 1 - e^{-8 \times .15} = 1 - e^{-1.2} = .699$.

(b) $\mu_T = 1/\lambda = 1/8$ min; the s.d. is also $1/\lambda = 1/8$ min.

(c) Looking backward in time one sees the same process as looking ahead.
The mean time to the most recent call is 1/8 min, and the mean time to
the next call is 1/8; the mean time between these two calls is 1/4 min, or
15 seconds. [This is true, despite the fact that looking ahead from the
point of that most recent call, the mean time to the next is 1/8 min.]

(d) The time to the 10th call is the sum of the times from one call to the
next; each of these is 1/8 min, so the mean time to the 10th is 10/8 min.
(e) The time to the 10th call exceeds 1 minute if and only if there are at
most 9 calls in that 1 minute period; look in the Poisson table for the
probability of at most 9, when the rate is 8: .717.

6-H **Solution:**

(a) The number arriving in a given interval is the sum of the numbers of the three types, and this is Poisson, with rate equal to the sum of the rates of the three types—30/hr. So the mean time to the next arrival is 1/30 of an hour, or 2 minutes; the mean time to the 10th arrival is 20 minutes.

(b) At the rate of 10 per hour ($\lambda = 10$), the average wait for one is $1/\lambda$, and for 5, it's 5 times that: $5/\lambda = 1/2$ hr.

(c) If we record only customers of types A and C, the process is Poisson with rate $\lambda = 10 + 5$, and mean time to an arrival equal to $1/\lambda = 1/15$ hr or 4 minutes.

(d) The simplest approach is to use the fact that over a long period of time, the ratio of type A arrivals to type B arrivals is 2:3. Selecting a time at random is equivalent to picking the next customer at random, so the probability that the next (among those that are A or B) is A is 2/5.

The long way around is to get the joint p.d.f. of U and V, the times to a type A and to a type B, respectively; this is the product of their individual p.d.f.'s:

$$f_{U,V}(u, v) = 10e^{-10u} \times 15e^{-15v}, \ u > 0, \ v > 0.$$

The probability that $U < V$ is the double integral of this p.d.f. over the region where this inequality holds:

$$P(U < V) = \iint\limits_{0 < u < v} 150\, e^{-10u - 15v} du\, dv = \tfrac{2}{5}.$$

6-I **Solution:**

(a) The integral looks like a gamma function, except for the 3 in front of the x. We can fix this by letting $3x = y$, so the integral becomes

$$\frac{1}{3^{14}} \int_0^\infty y^{13} e^{-y}\, dy = \frac{\Gamma(14)}{3^{14}} = \frac{13!}{3^{14}} = 1301.9.$$

(b) This is $\Gamma(1 + \tfrac{9}{2}) = \tfrac{9}{2} \cdot \Gamma(\tfrac{9}{2}) = \ \cdots \ = \tfrac{9}{2} \cdot \tfrac{7}{2} \cdot \tfrac{5}{2} \cdot \tfrac{3}{2} \cdot \tfrac{1}{2} \cdot \Gamma(\tfrac{1}{2}) = \tfrac{945}{32}\sqrt{\pi}.$

6-J The mean is defined as $\displaystyle\int x\, f(x \mid \theta)\, dx = \int_0^\infty \theta x^2 e^{-\theta x^2/2} dx$. Now set

$$u = \theta x^2/2, \ x = \sqrt{2u/\theta}, \text{ and } du = \theta x\, dx, \text{ or } dx = \frac{du}{\theta\sqrt{2u/\theta}}:$$

$$EX = \int_0^\infty 2u e^{-u}\frac{du}{\theta\sqrt{2u/\theta}} = \sqrt{\tfrac{2}{\theta}}\int_0^\infty u^{1/2} e^{-u}\, du = \sqrt{\tfrac{2}{\theta}}\,\Gamma(3/2) = \sqrt{\tfrac{\pi}{2\theta}}.$$

6-K   From Table IIb we find $z_{.90} = 1.2816$.

(a) $.90 = P\left(\dfrac{\chi^2 - 25}{\sqrt{50}} \le \dfrac{\chi^2_{.90} - 25}{\sqrt{50}}\right) \doteq \Phi(z_{.90})$, so $\chi^2_{.90} \doteq 25 + 1.2816\sqrt{50}$, or 34.06.

(b) From the footnote to Table Va, $\chi^2_{.90} = \frac{1}{2}(1.2816 + \sqrt{49}\,)^2 = 34.29$.

(This is closer to the 90th percentile 34.4 given in Table Va.)

6-L   **Solution:**
To find the mean, we integrate $1/x$ with respect to the p.d.f. of chi$^2(k)$:

$$E\left(\frac{1}{X}\right) = \int_{-\infty}^{\infty} \frac{1}{x} f(x)\,dx = \frac{1}{2^{k/2}\Gamma(k/2)} \int_0^{\infty} \frac{1}{x} \cdot x^{k/2-1} e^{-x/2}\,dx$$

$$= \frac{1}{2^{k/2}\Gamma(k/2)} \int_0^{\infty} x^{k/2-2} e^{-x/2}\,dx = \frac{2^{k/2-1}\Gamma(k/2-1)}{2^{k/2}\Gamma(k/2)} = \frac{1}{k-2}.$$

6-M   **Solution:**
The variables squared and summed are standard scores of the normally distributed X's, so they are standard normal. The sum of squares of 10 squares of standard normal variables has a chi$^2(10)$ distribution (by definition). The desired probability is therefore the tail-probability of χ^2 at 16, or (from Table Va) $P(\chi^2(10) > 16) = .100$.

6-N   **Solution:**
(a) The time to failure is Exp(1/2), so $P(L > 4) = e^{-4/2} \doteq .135$. (You could find this in Table IV, where $m = 2$, $c = 0$.)

(b) What is described is a parallel system, with reliability

$$R(4) = 1 - [1 - R_1(4)]^2 = 1 - (1 - e^{-2})^2 \doteq .252,$$

where R_1 and R_2 are both the reliability calculated in (a).

(c) The reliability as a function of t, calculated as in (b), is

$$R(t) = 1 - (1 - e^{-t/2})^2 = 2e^{-t/2} - e^{-t}.$$

The mean can be calculated as the integral of R [see (2) of §6.5]:

$$ET = \int_0^{\infty} (2e^{-t/2} - e^{-t})\,dt = 4 - 1 = 3.$$

A more intuitive approach (also correct) is to pick one unit, say unit 1, and wait until it fails. Then go to the other; by symmetry, it is alive with probability 1/2, and if alive, its mean time to failure is 2 hrs. Then

$$ET = E(T_1) + \tfrac{1}{2}E(T_2) = 2 + 1 = 3.$$

6-O   **Solution:**

For the system consisting of the first two blocks in parallel, the reliability function is $1 - (1 - R_1)(1 - R_2)$. A similar expression applies for the second two blocks. Putting these in series, we multiply to find the system reliability function:

$$R = [1 - (1 - R_1)(1 - R_2)] \cdot [1 - (1 - R_3)(1 - R_4)].$$

6-P   **Solution:**

(a) The probability to the left of 1 is the c.d.f. at 1:

$$P(T < 1) = F(1) = 1 - R(1) = 1 - 1/e = .632.$$

(b) The mean, from (2) of §6.5, is the integral of $R(t)$:

$$\int_0^\infty e^{-t^2} dt = \frac{\sqrt{\pi}}{2}.$$

CHAPTER 7: Additional Problems

Section 7.1

7-A   Consider these data:

	Mother		Father				Mother		Father			
Wt.	E	S	E	S	C	Wt.	E	S	E	S	C	
10.9	2	N	2	Q	1	5.6	2	S	3	Q	1	
6.1	2	S	2	S	1	6.6	4	N	4	S	6	
5.6	4	N	4	N	1	7.8	1	N	2	Q	1	Education (E)
7.2	2	Q	4	Q	1	7.4	2	Q	3	Q	1	4 = College grad
7.3	4	Q	2	Q	1	9.2	3	N	4	N	6	3 = Some college
7.9	4	N	4	N	6	7.6	3	N	4	N	1	2 = HS grad
6.6	4	S	2	S	1	6.6	1	S	1	S	2	1 = 8-12 grade
8.0	3	S	4	S	6	7.4	2	N	3	N	4	0 = < 8 grade
8.3	1	N	2	N	1	6.4	1	S	2	N	6	
8.5	3	N	4	N	4	8.3	0	N	1	N	1	
7.1	4	N	4	Q	6	5.4	1	S	2	Q	4	Smoking (S)
7.2	4	S	4	S	6	7.1	3	S	4	S	4	S = Smokes
7.1	3	N	4	N	1	9.9	4	S	4	N	6	Q = Quit
7.1	2	N	3	N	6	6.7	1	N	2	Q	3	N = Never
6.0	4	S	4	N	6	6.8	4	Q	4	N	6	
8.3	4	N	4	N	5	5.9	1	S	1	N	2	
7.8	2	S	2	S	3	7.1	1	S	2	S	4	Church (C)
6.9	2	S	2	N	1	6.5	2	N	4	N	1	1 = All attend
7.8	2	N	2	N	1	6.7	1	N	1	Q	1	2 = Mother and
7.9	3	N	2	S	1	6.8	2	Q	4	Q	1	children attend
5.4	3	Q	2	N	1	6.9	2	S	3	S	4	3 = Only children
6.1	4	N	4	N	6	3.3	4	N	4	N	4	attend
5.6	2	N	0	S	1	7.5	2	N	0	N	1	4 = Sporadic
7.8	2	N	4	Q	1	5.3	2	N	0	N	1	5 = Holy days
8.4	3	N	4	Q	6	6.6	2	N	2	N	1	6 = None
7.4	4	N	4	Q	1	6.1	4	N	1	Q	1	
8.6	4	N	4	Q	2	7.6	2	S	1	S	6	
8.8	3	N	4	N	6	7.6	3	S	1	Q	4	
9.2	4	N	4	S	1	7.3	4	N	4	Q	6	
6.6	4	N	4	Q	1	5.6	2	N	4	N	2	
6.6	1	N	2	N	1	7.1	2	N	2	S	4	
6.1	4	N	4	Q	1	6.6	4	Q	4	N	6	
6.6	2	S	4	N	6	6.9	4	N	4	N	4	
6.9	3	Q	3	Q	6	7.8	2	N	3	Q	1	
4.5	2	S	1	Q	1	8.6	0	N	3	S	1	
7.8	4	N	4	Q	1	8.6	4	N	4	N	6	

The data[1] are for 72 couples enrolled in an HMO, taken at the time of the birth of their child: W is the birth-weight of the child (oz); E is the education and S is the smoking status, for the father and for the mother; C denotes church-going habits of the family.

(a) Make a cross-classification on the three variables C, S of father, and S of mother, using mother's smoking status as the layering variable.

(b) Obtain a 2-way classification of father's and mother's smoking status from (a).

(c) In what fraction of those couples in which the mothers smoke do the fathers smoke?

(d) Construct the contingency table for education of the father and of the mother. (Comment?)

Sections 7.2-7.4

7-B Make a stem-leaf diagram or other convenient display of the weight data in Problem 7-A.

7-C Use the stem-leaf diagram in 7-B to find the five-number summary needed for a box-plot of the weight data, and construct the box-plot.

7-D Tests of a process for removing nitrogen oxides from flue gases yielded the following data on the amount removed (in %) from a coal-burning facility in Pittsburgh:[2]

  91 95 90 83 91 65 55 42 55 81 89 38 20 45 58 85 78 70

Construct a stem-leaf diagram, and find the median, mean, quartiles, range, interquartile range, and midrange.

Section 7.5

7-E Make a frequency table for the birth-weight data in Problem 7-A, using as interval midpoints 3.4, 4.1, 4.8, 5.5, etc., and find the mean and s.d. after introducing the coding variable $Y = (X - 6.9)/.7$, to find first the mean and s.d. of the Y's.

7-F Show that the sample standard deviation is larger than the sample mean deviation (when $n > 1$).

---

[1]Extracted from an extensive data set given in Hodges, Krech, and Crutchfield, *Stat Lab* (McGraw-Hill, 1975). The ages of the mothers represented in Table 7.10 were 31-33 years.

[2]J. T. Yeh, et al., *Environmental Progress* (1985), 223-228.

7-G Using a statistical hand calculator, enter these numbers: 900,001, 900,002, 900,003, and find S. This should be the same as the s.d. of the numbers 1, 2, 3 (why?). Is it? Can you find the s.d. in your head?

7-H Given $n = 71$, $S = 1.149$, and $\sum X_i = 510.3$, find $\sum X_i^2$.

Section 7.6

7-I Mineral compositions were determined for 12 rocks taken from an area in England with many excellently preserved plant fossils.[3] Quartz and siderite concentrations (in %) were reported as follows:

Rock	Quartz	Siderite
1	34.3	62.2
2	37.5	58.3
3	22.3	74.7
4	18.7	75.0
5	27.9	64.9
6	36.7	59.0
7	19.8	74.3
8	21.6	11.1
9	7.9	89.5
10	31.5	58.3
11	7.0	74.7
12	24.4	63.5

A plot quickly reveals that #8 is quite different from the rest. (One possibility is that there was an error in measuring or recording it. Yet another possibility is that the measurement is correct, but that this rock was formed during a different geological period.) Calculate the correlation coefficient with and without observation #8, given the following sums for all 12 cases:

$$\sum X = 289.6, \quad \sum Y = 765.5, \quad \sum XY = 17{,}690.73,$$
$$\sum X^2 = 8{,}112.64, \quad \sum Y^2 = 52{,}831.01.$$

7-J Suppose a weather forecaster's predictions of the high temperature for the following day, for a period of one month, have a correlation coefficient of .85 when the data are in degrees Fahrenheit. What is the correlation coefficient when the data are converted to degrees Celsius? $(F = 32 + \frac{9}{5}C)$

---

[3]C. R. Hill, D. T. Moore, J. T. Greensmith, and R. Williams, "Palaeobotany and petrology of a middle Jurassic ironstone bed at Wrack Hills, North Yorkshire," *Yorkshire Geological Society* **45** (1985).

Chapter 7: Solutions

7-A   Solution:

(a) The 2-way tables in the 3 layers are as follows:

Mother never smoked:

Church:

	1	2	3	4	5	6	
N	11	1	0	4	1	6	23
Q	10	1	1	0	0	3	15
S	4	0	0	1	0	1	6

Father's smoking:

Mother quit smoking:

Church:

	1	2	3	4	5	6	
N	1	0	0	0	0	2	3
Q	4	0	0	0	0	1	5
S	0	0	0	0	0	0	0

Father's smoking:

Mother smokes:

Church:

	1	2	3	4	5	6	
N	1	1	0	0	0	4	6
Q	2	0	0	2	0	0	4
S	2	1	1	3	0	3	10

Father's smoking:

(b) This table is constructed using the marginal totals shown in (a):

Mother:

	N	Q	S
N	23	3	6
Q	15	5	4
S	6	0	10

Father:

(c) Of the 20 ($= 6 + 4 + 10$) mothers who smoke, 10 of their husbands also smoke: 10/20.

(d)

		Mother's education:					
		0	1	2	3	4	
	0	0	0	2	0	0	2
	1	1	3	2	1	1	8
Father's education:	2	0	8	7	2	2	18
	3	1	0	5	1	0	7
	4	0	0	6	10	21	37
		2	10	22	14	24	72

[Our comment: The levels agree (main diagonal) in nearly half of the couples; and in all but 13, they are within one level of each other (i.e., on the three middle diagonals).]

7-B Solution:

Most of the weights begin with 6 or 7, so splitting the stems should help in a stem-leaf display:

Depth	Stm	Leaves
1	3	1
1	3	
1	4	
2	4	5
5	5	443
10	5	66669
16	6	011114
33	6	56666666677889999
13	7	1111112233444
26	7	566688888899
14	8	033345
8	8	5668
4	9	22
2	9	9
1	10	
1	10	9

$n = 72$ (bracketing the 33, 13, 26 rows)

Class	Freq.
3.1-3.7	1
3.8-4.4	0
4.5-5.1	1
5.2-5.8	7
5.9-6.5	8
6.6-7.2	24
7.3-7.9	17
8.0-8.6	9
8.7-9.3	3
9.4-10.0	1
10.1-10.7	0
10.8-11.4	1

(This is only one of many possible frequency distributions, depending on the class-interval scheme.)

7-C Solution:

There are 72 observations in all, so the median lies between the 36th and 37th smallest, or 7.1—the average of the 3rd and 4th (in order) on the 7-stem. The 1st quartile is between the 18th and 19th smallest or 6.6. (Ideally, there would be 18 smaller and 54 larger, but because of duplications, this isn't quite the case.) Similarly, we take 7.7 as Q_3, the average of the 18th and 19th counting from the largest. The box plot is shown on the next page.

Box plot for 7-C:

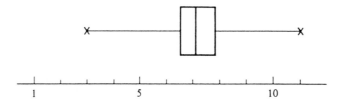

7-D   Solution:

With just 2 digits, the first is the stem and the second the leaf:

Depth	Stm	Leaves
1	2	0
2	3	8
4	4	25
7	5	558
8	6	5
2	7	08
8	8	1359
4	9	0115

The median is on the 7-stem, between the 9th and 10th smallest: $\tilde{X} = 74$.
The midrange is half-way from the smallest to the largest: $\frac{1}{2}(20 + 95)$, or
57.5. The range is the width of the interval from 20 to 95: $R = 75$. If
we take Q_1 to be the median of the lower 9 observations, it is 55. The
median of the upper 9 is $Q_3 = 89$, so $\text{IQR} = Q_3 - Q_1 = 34$.

The sample sum is 1231, so $\bar{X} = 68.4$.

7-E   Solution:

x_i	f_i	y_i	$f_i y_i$	$f_i y_i^2$
3.4	1	-5	-5	25
4.8	1	-3	-3	9
5.5	7	-2	-14	28
6.2	8	-1	-8	8
6.9	24	0	0	0
7.6	17	1	17	17
8.3	9	2	18	36
9.0	3	3	9	27
9.7	1	4	4	16
10.4	0	5	0	0
11.1	1	6	6	36
	72		24	202

$X_i = 6.9 + .7 Y_i$, so $\bar{X} = 6.9 + .7 \times \frac{24}{72} \doteq 7.13$

$S_Y^2 = \frac{1}{71}[202 - 72 \times (\frac{1}{3})^2] \doteq 2.732$,

$S_Y = 1.653$, so $S_X = .7 \times 1.653 \doteq 1.16$.

[For the ungrouped data, $\bar{X} = 7.13$, and
$S = 1.22$.]

7-F **Solution:**
We know that the s.d. with divisor n, called $\sqrt{V}$ in the text, is at least as large as the m.a.d.—as a special case of what is true for distributions generally (see Problem 3-22, or §5.6). That S, the version with divisor $n - 1$ is actually larger than the m.a.d. is seen as follows:

$$S = \sqrt{S^2} = \sqrt{\frac{n}{n-1}V} > \sqrt{V} \geq \text{m.a.d.}$$

7-G **Solution:**
What you get depends on the particular calculator you use. Most will give 0 or an error message, but the correct answer is 1. (In your head: the deviations from the mean of 1 are -1, 0, 1, with squares 1, 0, 1, and dividing the sum of squares by $n - 1 = 2$ gives 1 as the variance.) The deviations are the same, with and without the 900,000 tacked on, so the s.d. is the same.

7-H **Solution:**
Solving the formula for S^2 for the sum of squares yields

$$\sum X_i^2 = (n-1)S^2 + \left(\sum X_i\right)/n = 70 \times 1.149^2 + 510.3^2/71 = 3760.$$

7-I The scatter diagram is shown in Figure 8:

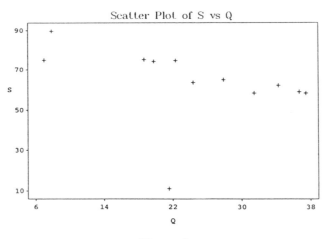

Figure 8

Substituting in the formula for the correlation coefficient, we obtain

$$r = \frac{17690.73 - (289.6)(765.5)/12}{\sqrt{8112.64 - (289.6)^2/12}\sqrt{52831.01 - 765.5^2/12}} = -.3696.$$

With #8 deleted, the calculation results in $r = -.885$. That point is certainly influential. Observe that except for #8, the points are indeed close to being on a line, and we can see from the data that the sum $x + y$ is roughly constant. This suggests that quartz and siderite together make up some nearly fixed proportion of the rocks—except for #8.

7-J   Solution:

The correlation is the same, since the conversion is a linear one.

CHAPTER 8: Additional Problems

Sections 8.1-8.2

8-A   Consider a random sample of size 3 from a population with p.d.f.
$f(x \mid \alpha) = \alpha x^{\alpha - 1}$ for $0 < x < 1$, where α is a nonnegative parameter.
[This includes $\mathcal{U}(0,\, 1)$ as the special case where $\alpha = 1$.]
(a) Write the joint p.d.f. of the observations.
(b) Write the likelihood function for this sample of size 3: $(\frac{1}{2}, \frac{2}{3}, \frac{3}{5})$.

8-B   Find the joint probability function of the observations in

(a) a random sample from Poi(m): $f(x \mid m) = \dfrac{m^x}{x!} e^{-m}$.

(b) a simple random sample of size n from a Bernoulli population of
size N, where $P(1) = p = M/N$, $P(0) = 1 - p$. ("Simple random sample"
means that observations are drawn at random, but without replacement.)

8-C   Give the likelihood functions (stripped of unnecessary constants):
(a) $L(m)$ for the sample in 8-B(a).
(b) $L(M)$ for the sample in 8-B(b).

8-D   Suppose U has the Maxwell density:
$$f(u \mid \sigma) = \sqrt{\frac{2}{\pi}} \cdot \frac{u^2}{\sigma^3} \exp\left(-\frac{u^2}{2\sigma^2}\right), \ \ u > 0.$$
(a) Show that this is a member of the exponential family.
(b) Give the likelihood function, for a random sample of size n.

Section 8.3

8-E   Find a sufficient statistic, given a random sample, for the parameter in the
(a) Maxwell distribution defined in Problem 8-D.
(b) in a Rayleigh distribution, defined by the p.d.f. (for $\theta > 0$)
$$f(\theta \mid x) = \theta x \exp\left(-\tfrac{1}{2}\theta x^2\right), \ \ x > 0.$$

8-F   Let $\mathbf{X} = (X_1, \ldots, X_n)$ be a sample drawn at random without replacement
from a Bernoulli population of size N, with $p = M/N$. Find the conditional p.f. of $\mathbf{X}$ given $\sum X_i = k$.

Sections 8.4–8.5

8-G A bowl contains eight chips, numbered 1 through 8. Find the distribution of the *median* of the numbers in a random selection of 3 chips (without replacement).

8-H Find the sampling distribution of $\sum X^2$, where X is a random sample of size n from $N(0, 1)$.

8-I Explain how to simulate a random sample from a Cauchy distribution with median 0, given a table or other means of generating random digits.

Section 8.6

8-J Find the probability that the smallest observation in a random sample of size 10 from $U(0, 1)$ is at least .15.

8-K Find the mean of the largest observation in a sample of size $n = 3$ from the population in Problem 8-A: $f(x \mid \alpha) = \alpha x^{\alpha - 1}$ for $0 < x < 1$.

8-L Find the covariance of the smallest and largest observations in a random sample of size n from $U(0, 1)$.

8-M Find the mean and variance of the sample range, for random samples of size 5 from $U(0, 1)$.

8-N Find the expected value of the median of a random sample from a population that is symmetric about the value $x = a$.

Sections 8.7–8.10

8-O In one year recently, the average ACT score in Minnesota was 19.4. If we assume a standard deviation of 2.0, find the probability that the mean of a random sample of 25 scores from Minnesota exceeds 20.0.

8-P A company packages paper clips in boxes labeled "100 clips." In counting clips in these boxes we've found that the number in a box varies from as low as 94 to well over 100. Suppose that the mean is actually 100, and the standard deviation is 2.0. Find the probability that a carton of 36 boxes contains fewer than 3580 clips in all (more than 20 shy of 3600).

8-Q A machine screw may be skipped or damaged in the threading process. If .5% of the screws produced by a particular machine are defective, find the probability that more than 2 screws in a box of 200 are defective.

8-R In Problem 5-AA, we found the m.g.f. of X with p.d.f. $f(x) = \frac{1}{2}e^{-|x|}$ to be $(1 - t^2)^{-1}$. The mean is 0 and the variance is 2. For a random sample of size n, find the m.g.f. of the standardized variable $Z = \sum X_i / \sqrt{2n}$, and show that it converges to the m.g.f. of $\mathcal{N}(0, 1)$.

Section 8.11-8.12

8-S Consider inference about the parameter θ in a Rayleigh distribution, defined by the p.d.f.

$$f(\theta \mid x) = \theta x \exp\left(-\tfrac{1}{2}\theta x^2\right), \quad x > 0,$$

where θ is a positive parameter. Find a conjugate prior, and the posterior distribution that results from its use.

8-T Show that if the distribution of data X does not depend on the parameter θ, the posterior for θ is the same as the prior.

8-U You are a military intelligence expert, sent on a secret mission to an enemy training area to count enemy tanks. You know that the enemy (not too wisely) numbered its tanks in the area from 1 to N. You observe, secretly, tanks numbered 2, 7, 13, 5, 8. Assume a uniform prior distribution for N and find the posterior.

8-V Consider a random variable X with two possible values, and a family of just three possible models, defined by this table of probabilities:

	θ_1	θ_2	θ_3
a	.2	.5	.9
b	.8	.5	.1

(The columns of this table define the models; the rows define the likelihoods.) Assume a prior distribution (2/4, 1/4, 1/4) for $(\theta_1, \theta_2, \theta_3)$, respectively and determine the posterior distribution
(a) after observing $X = a$.
(b) after $X = a$ in each of two independent trials, using the posterior from part (a), as a new prior for the second observation.
(c) after $X = a$ in each of two independent trials, applying the original prior to a table of distributions for the data pair (a, a).

Chapter 8: Solutions

8-A   Solution:

In each case, the observations are independent, so their joint p.d.f. is the product of the marginal p.d.f.'s.

(a) $f(x_1, x_2, x_3 \mid \theta) = f(x_1 \mid \alpha)f(x_2 \mid \alpha)f(x_3 \mid \alpha) = \alpha^3 x_1 x_2 x_3,\ 0 < x_i < 1$ for $i = 1, 2, 3$ (and 0 elsewhere).

(b) The likelihood function is given by the expression for the p.d.f. in (a); substituting the observed data for the x's (and dropping 5^{-1}), we get

$$L(\alpha) = \alpha^3 (\tfrac{1}{2} \cdot \tfrac{2}{3} \cdot \tfrac{3}{5})^\alpha = \frac{\alpha^3}{5^\alpha}.$$

8-B   Solution:

(a) Again, a product: $f(x_1, ..., x_n \mid m) = \prod_1^n \dfrac{m^{x_i}}{x_i!} e^{-m} = \dfrac{m^{\Sigma x_i}}{\Pi x_i!} e^{-nm}.$

(b) The observations are not independent. The joint p.f. of the observations in the sample sequence was given in Chapter 4 [as (4) in §4.1], so we just repeat it here:

$$f(x_1, ..., x_n \mid m) = \frac{\binom{M}{\Sigma x_i}\binom{N-M}{n-\Sigma x_i}}{\binom{n}{\Sigma x_i}\binom{N}{n}},\quad x_i = 0 \text{ or } 1,\quad i = 1, ..., n.$$

8-C   Solution:

(a) Since the x_i's are fixed, we can drop the product in the denominator:

$$L(m) = m^{\Sigma x_i} e^{-nm},\ m > 0.$$

(b) There is no parameter M in the denominator, so the likelihood is

$$L(M) = \binom{M}{\Sigma x_i}\binom{N-M}{n-\Sigma x_i},\ M = 1, 2, , ..., N.$$

8-D   Solution:

(a) We show this by identifying the functions B, h, R, Q:

$$B(\sigma) = \sigma^3,\ h(u) = u^2,\ R(u) = u^2,\ Q(\sigma) = -\frac{1}{2\sigma^2}.$$

(b) The constant and the u_i's outside the exponential can be left out when writing the product of the $f(u_i)$'s:

$$L(\sigma) = \sigma^{-3n} \exp\left(-\sum_1^n \frac{u_i^2}{2\sigma^2}\right).$$

8-E   **Solution:**

(a) The likelihood function is given in the solution to 8-D (just above). It depends on the data through the value of the sum of squares of the sample observations: $\sum U_i^2$ is sufficient.

(b) The likelihood function is $\theta^n \exp\{-\frac{1}{2}\sum X_i^2\}$. Since it depends on the data through the value of $\sum X_i^2$, that sum is sufficient.

8-F   **Solution:**

When $\mathbf{X} = (X_1, ..., X_n)$ is a sample drawn at random without replacement from a finite Bernoulli population, the joint p.f. of the (unconditional) distribution of the X's is

$$f(x_1, ..., x_n \mid M) = \frac{\binom{N}{\sum X_i}\binom{N-M}{n-\sum X_i}}{\binom{n}{\sum X_i}\binom{N}{n}}, \quad x_i = 0, 1.$$

The distribution of the sample sum is hypergeometric, with p.f.

$$f(k \mid M) = \frac{\binom{N}{k}\binom{N-M}{n-k}}{\binom{N}{n}}.$$

To find the conditional p.f., we use the definition of conditional probability, which we write using boldface type for the sample vector:

$$P(\mathbf{X} = \mathbf{x} \mid \sum X_i = k) = \frac{P(\mathbf{X} = \mathbf{x} \text{ and } \sum X_i = k)}{P(\sum X_i = k)}.$$

The event in the numerator of the last fraction is empty unless the x_i add up to k. But if they *do* add up to k, then $\sum X_i$ equals k, and it is not necessary to repeat this:

$$P(\mathbf{X} = \mathbf{x} \mid \sum X_i = k) = \frac{P(\mathbf{X} = \mathbf{x})}{P(\sum X_i = k)} = \frac{f(x_1, ..., x_n \mid M)}{f(k \mid M)},$$

the ratio of the p.f.'s given above. Since this holds when $\sum X_i = k$, everything cancels out except the first factor in the denominator of $f(x_1, ..., x_n \mid M)$, leaving just

$$P(\mathbf{X} = \mathbf{x} \mid \sum X_i = t) = \frac{1}{\binom{n}{k}}, \quad \text{if } \sum x_i = k,$$

and (as explained above) 0, otherwise. [The significance of the fact that this conditional p.f. is independent of M is explained at the end of §8.3.]

8-G   **Solution:**

There are $\binom{8}{3} = 56$ equally likely samples. Those with median 2 are as follows: 123, 124, 125, 126, 127, 128—six in number. Those with median 3 can start either with 1: 134, 135, 136, 137, 138, or with 2: 234, 235, 236, 237, 238—ten in number. Similarly there are 12 samples with median 4, 12 with median 5, 10 with median 6, and six with median 7. (Having done the count for 2, 3, and 4, one should not find it necessary to count for 5, 6, and 7, because of the population symmetry.) Dividing each count by 56 produces the p.f. of the sample median:

k	2	3	4	5	6	7
$P(\tilde{X} = k)$	6/56	10/56	12/56	12/56	10/56	6/56

8-H   Solution:

Since the X's are standard normal, this is the sum of squares of independent standard normal variables, which defines chi-square: $\sum_i X_i^2 \sim \text{chi}^2(n)$.

8-I   Solution:

Given $U \sim \mathcal{U}(0, 1)$, $X = F^{-1}(U)$ has the c.d.f. F [see §8.5]. To create variables with the Cauchy c.d.f., we find the inverse of F. The c.d.f. is

$$F(x) = \tfrac{1}{\pi}[\tfrac{\pi}{2} + \text{Arctan } x] = u.$$

Solving for u, we find

$$F^{-1}(u) = \tan(\pi u - \tfrac{\pi}{2}).$$

So, given a random sample $(U_1, ..., U_n)$ from $\mathcal{U}(0, 1)$, as in Table XV, define

$$X_i = \tan(\pi U_i - \tfrac{\pi}{2}).$$

These constitute a random sample from the population with c.d.f. F.

For instance, reading from the bottom line of the first page of Table XV, we have .514, .457, .083, .967, .270, .209, .380, .655, .653, .518 (with a decimal in front of the group of integers, and rounding to 3 places). The X's are $\tan(.514\pi - \pi/2) = .044$, $\tan(.457\pi - \pi/2) = -.136$, and so on:

.044, $-.136$, -3.75, 9.61, $-.882$, -1.30, $-.396$, .535, .521, .057.

8-J   Solution:

The smallest observation is at least .15 if and only if every observation is at least .15:

$$P(X_{(1)} > .15) = P(X_1 > .15, \cdots, X_{10} > .15) = (.85)^{10} = .197.$$

8-K   Solution:

The population c.d.f. is $F(x) = x^\alpha$, $0 < x < 1$, so the p.d.f. of Y, the largest in a sample of 3 is

$$f_Y(y) = 3(y^\alpha)^2 \cdot \alpha y^{\alpha-1} = 3\alpha y^{3\alpha-1}.$$

And the mean:

$$EY = 3\alpha \int_0^1 y^{3\alpha} dy = \frac{3\alpha}{3\alpha + 1}.$$

8-L Solution:

From (6) in §8.6, the joint p.d.f. of the smallest (U) and largest (V) is

$$n(n-1)(v-u)^{n-2}, \ 0 < u < v.$$

The expected product is

$$E(UV) = n(n-1) \int_0^1 v \left\{ \int_0^v u(v-u)^{n-2} du \right\} dv.$$

In the inner integral, let $u = v - vw$:

$$\int_0^v (v-u)^{n-2} du = v^n \int_0^1 w^{n-2}(1-w) dw = \frac{v^n}{n(n-1)}.$$

Then,

$$E(UV) = \int_0^1 v^{n+1} dv = \frac{1}{n+2},$$

and

$$\sigma_{U,V} = E(UV) - EU \cdot EV = \frac{1}{n+2} - \frac{1}{n+1} \cdot \frac{n}{n+1} = \frac{1}{(n+2)(n+1)^2}.$$

(The means, EU and EV, were obtained in Example 8.6b.)

8-M Solution:

The p.d.f. of the sample range R is given in Example 8.6c; when $n = 5$,

$$f_R(r) = 20r^3(1-r), \ 0 < r < 1.$$

To find the kth moment, we integrate:

$$E(R^k) = 20 \int_0^1 r^{3+k}(1-r)\, dr = 20 \left\{ \frac{1}{4+k} - \frac{1}{5+k} \right\} = \frac{1}{(4+k)(5+k)}.$$

So,

$$ER = \frac{1}{30}, \ E(R^2) = \frac{1}{42}, \ \text{var } R = \frac{1}{42} - \frac{1}{900} = .0227.$$

The s.d. is about .15.

8-N Solution:

From Problem 8-2 we know that the distribution of the sample median is also symmetric about $x = a$. If the mean value exists, it is $EX = a$.

8-O Solution:

From (4) and (5) of §8.7, $E(\overline{X}) = \mu = 19.4$, and $\sigma_{\overline{X}} = 2.0/\sqrt{25} = .40$. Using (4) of §8.8, we have

$$P(\overline{X} > 20) = 1 - F_{\overline{X}}(20) \doteq 1 - \Phi\left(\frac{20 - 19.4}{.4}\right) = \Phi(-1.5) = .0668.$$

8-P   Solution:

The population variable is discrete (a count), and we don't know the shape of the distribution, but the central limit theorem makes it unnecessary to know that shape. Given $\mu = 100$ and $\sigma = 2.0$, the number in 36 boxes is 3600 on average, and the s.d. of the number is $\sigma \times \sqrt{36} = 12$. Then

$$P(\text{total} < 3580) \doteq \Phi\left(\frac{3580 - 3600}{12}\right) = \Phi(-5/3) \doteq .048.$$

8-Q   Solution:

The number in a box of 200 is Bin(200, .005), but since $npq = .995 < 5$, the normal approximation is not as good as the Poisson. The Poisson parameter is $np = 1$, so

$$P(\text{more than 2 def.}) = 1 - e^{-1}(1 + 1 + \tfrac{1}{2}) = .0803.$$

8-R   Solution:

The m.g.f. of Z is

$$E\{e^{t(\Sigma X_i/\sqrt{2n})}\} = \left\{E(e^{tX_i/\sqrt{2n}})\right\}^n = \psi_X^n(t/\sqrt{2n}) = \left(\frac{1}{1 - t^2/(2n)}\right)^n.$$

Now let $-t^2/(2n) = y$, so that the m.g.f. of Z becomes

$$[(1+y)^{1/y}]^{-t^2/2}.$$

The quantity in brackets tends to e as n becomes infinite (and $y \to 0$), so the m.g.f. converges to $e^{-t^2/2}$, the m.g.f. of a standard normal variable.

8-S   Solution:

The likelihood function is $L(\theta) = \theta^n \exp(-\tfrac{1}{2}\theta \sum X_i^2)$, which shows that the sum of squares is sufficient. What's called for is a family of distributions for θ whose p.d.f.'s would combine nicely with $L(\theta)$. Gamma distributions should do the trick:

$$g(\theta) \propto \theta^{\alpha - 1} e^{-\lambda\theta}, \quad \theta > 0.$$

With this prior, the posterior is

$$h(\theta) \propto g(\theta) h(\theta \mid \mathbf{X}) \propto \theta^{n+\alpha-1} e^{-\theta(\lambda + \Sigma X_i^2/2)}.$$

This is also a gamma density: $\text{Gam}(n + \alpha, \lambda + \sum X_i^2/2)$. So the family of gamma distributions is a conjugate family for Rayleigh densities.

8-T   Solution:

In this case, the likelihood function is independent of θ—as a function of θ, it is a constant. Hence, the posterior is just proportional to the prior:

$$h(\theta \mid x) \propto g(\theta) f(x \mid \theta) \propto g(\theta),$$

which means that the prior and posterior are the same distribution.

8-U　**Solution:**

The number of positive integers is infinite, so we cannot really assign equal probabilities to them all. However, suppose we simply assign a constant weight, say 1, to each integer, and see where treating these as though they were probabilities leads. The likelihood function [Problem 8-15], is

$$L(N) = \begin{cases} 1/N^5, & N \geq X_{(5)}, \\ 0, & \text{otherwise.} \end{cases}$$

Formal application of "posterior $\propto$ likelihood $\times$ prior" gives us

$$h(N \mid X_{(5)} = 13) \propto \frac{1}{N^5}, \quad N = 13, 14, \dots.$$

This defines a proper distribution as the posterior, even though the prior is improper. (But it is what we'd get if we took a sequence of proper priors converging to the improper one we've used, and passed to the limit.) To find the p.f. of the posterior, we'd need to divide $1/N^5$ by the sum of such terms from 13 to ∞. (This can be done numerically—the sum is approximately 1.05×10^{-5}. The first few values of the p.f. are .257, .177, .125, .091, for $N = 13, 14, 15, 16$.)

8-V　**Solution:**

(a) The (unconditional) probability of $X = a$ is $\frac{.4}{4} + \frac{.5}{4} + \frac{.9}{4} = \frac{1.8}{4}$, and the posterior probabilities for $\theta_1, \theta_2, \theta_3$ are proportional to the terms in this sum: 4/18, 5/18, 9/18.

(b) Applying the posterior probabilities in (a) to the probabilities for $X = a$ in the given table, we obtain (new) posterior probabilities proportional to $4 \times .2$, $5 \times .5$, $9 \times .9$; they are 8/114, 25/114, 81/114.

(c) The three distributions for (X_1, X_2) are as follows:

	θ_1	θ_2	θ_3
aa	.04	.25	.81
ab	.16	.25	.09
ba	.16	.25	.09
bb	.64	.25	.01

Applying the original prior, and using the row for the data (a, a)—the first row, we get posterior probabilities proportional to

$$2 \times .04, \ 1 \times .25, \ 1 \times .81,$$

the same posterior as obtained step-wise in (a) and (b).

CHAPTER 9: Additional Problems

Sections 9.1-9.2

9-A   In Example 9.9c are given the following measurements of tracheal
compliance in newborn lambs: .029, .043, .022, .012, .020, .034.[4]
(a) Find the standard error of the mean.
(b) Suppose the investigators wanted to achieve a standard error of
at most .0020. Assuming S stays at about .011, about how many lambs
would be required?

9-B   In the data of Problem 7-A, we see that 20 out of 72 families reported no
church attendance at all. Give an estimate of the proportion of families
in the sampled population, together with a standard error.

9-C   Suppose we want to estimate the proportion of students on a campus who
had used alcohol in the last month.
(a) How large a random sample is required if we want the standard error
of our estimate not to exceed .05?
(b) If we are certain that the population proportion is at least .8, what
would be the minimum sample size needed to meet the same condition as
in (a)?

9-D   When $X \sim \mathcal{U}(0, \theta)$, the parameter θ is twice the mean: $\theta = 2\mu$. In
Example 9.1a, we considered the largest sample observation as an
estimator of θ. Suppose instead we use $2\bar{X}$ as an estimator of θ. Find
its bias and mean squared error.

Sections 9.3-9.4

9-E   Show that the sample proportion $\hat{p}$ is consistent in estimating the corresponding population proportion p, under random sampling, by calculating
the probability in an interval of width ϵ about p.

---

[4]V. K. Bhutani, R. J. Koslo, and T. H. Shaffer, "The effect of tracheal smooth
muscle tone on neonatal airway collapsability," *Peciatric Research* **20** (1986), 492-495.

9-F Durations (in minutes) of 74 storms in the Tampa Bay area are reported in a 1984 study.[5] The mean is 75.7 and the standard deviation 60.7. Assuming random sampling, give 95% confidence limits for the mean duration.

9-G A newspaper headline proclaimed "Most U.S. Catholics want changes in sexual policies." The article was a report of a 1986 Gallup Poll based on telephone interviews with 264 Roman Catholics, and the proportion favoring change in its church's policies was 57%. The fine print at the end of the article said that "19 times out of 20, the error attributable to sampling and other random effects should not exceed 8 percentage points in either direction." Explain, and give 95% confidence limits for the proportion of all Catholics who want change. (The sampling was said to be conducted at "scientifically selected places nationwide," but the sampling scheme was not otherwise described.)

Sections 9-5–9.7

9-H A random sample of 15 cigarettes of a certain brand was tested for nicotine content. The average was found to be 20.3 mg, and the s.d. of the content in the 15 cigarettes was 3.0 mg. Construct a 95% confidence interval for the mean content for all cigarettes of this brand. (Assume a random sample from a nearly normal population.)

9-I For a random sample from a population with p.d.f. $\lambda e^{-\lambda x}$, $x > 0$, show that the quantity $Y = \lambda \sum X$ is pivotal. [Hint: $\sum X \sim \text{Gam}(n, \lambda)$.]

9-J Show that the p.d.f. of the t-distribution [(5) in §9.6] tends to the p.d.f. of a standard normal variable (ignoring constant factors).

9-K Show that in the case of 1 d.f., the t-distribution is a Cauchy distribution with median 0.

9-L Consider a statistic T to be used for estimating a parameter θ. If the p.d.f. of T is $f_T(t \mid \theta)$, and the corresponding c.d.f., $F_T(t \mid \theta)$, show that the variable $V = F_T(T \mid \theta)$ is pivotal.

9-M Suppose you have computer software that gives percentiles of gamma distributions. Use the result of Problem 9-I to derive confidence limits for the parameter λ, given a random sample from $\text{Exp}(\lambda)$.

---

[5]"Lightning phenomenology in the Tampa Bay area," *J. Geophys. Res.* (1984), 11789-11805.

Sections 9.8-9.9

9-N   A research study concerned with air pollutants in the Lincoln Tunnel
under the Hudson River reported these summary statistics on concentrations in parts per million:[6]

	n	Mean	s.d.
1970	28	65.8	14.9
1983	28	15.6	3.8

(a) Give an estimate of the mean difference in concentration along with
the standard error of the estimate.
(b) Find 95% confidence limits for the mean difference.

9-O   Independent polls before ($n = 1200$) and after ($n = 950$) a particular
crisis show a drop in approval of the president's performance from 65%
before to 48% after. Find the standard error of an estimate of the change
in population proportion and construct 95% confidence limits for that
proportion.

9-P   The usual estimators for a variance of a normal population are multiples
of squared deviations about the mean:

$$Y = K \sum_{1}^{n} (X_i - \bar{X})^2 = K(n-1)S^2.$$

Find the multiplier K that minimizes the mean squared error.

9-Q   An experiment was conducted to check the variability in explosion times
of detonators.[7] Variability must be controlled to achieve simultaneity of
explosions. Data for run #1 are as follows, in milliseconds short of 2.7
seconds: 11, 23, 25, 9, 2, 6, −2, 2, −6, 8, 9, 19, 0, 2. Obtain 90%
confidence limits for the population s.d. σ.

Section 9.10

9-R   Given a random sample of size n from a population with the p.d.f. of
Problem 8-A: $f(x \mid \alpha) = \alpha x^{\alpha - 1}$ for $0 < x < 1$, obtain an estimator for α
(a) using the method of maximum likelihood.
(b) using the method of moments.

---

[6]"Non-methane organic composition in the Lincoln Tunnel," *Envir. Sci. &
Technology* **20** (1986).

[7]Reported in G. L. Tietjen and M. E. Johnson, "Exact statistical tolerance limits for
sample variances," *Technometrics* **21** (1979), 107–110.

9-S Example 8.2e obtained the likelihood function for θ, based on a random sample of size n from $\mathcal{U}(0, 1)$ as $L(\theta) = 1/\theta^n$ for $\theta > X_{(n)}$ (and 0 elsewhere).
(a) What is the m.l.e. of θ?
(b) Find the method of moments estimator.

Section 9.11

9-T Consider a random sample from $f(x \mid \theta) = \theta x e^{-\theta x^2/2}$, $x > 0$, where θ is a positive parameter. Assuming a quadratic loss function, find the Bayes estimate of θ given a gamma prior: $g(\theta) \propto \theta^{\alpha-1} e^{-\lambda \theta}$.

9-U A patient needs an organ transplant; in order to judge histocompatibility with possible donors, the patient's human leukocyte antigen (HLA) is determined. National blood bank records show that in a random sample of 5000 individuals, only one has the proper HLA type. Let λ denote the population rate per 5000 for this type, and assume a uniform (improper) prior for λ: $g(\lambda) = 1$. Find a 95% probability interval for λ as given by the 2.5 and 97.5 percentiles of the posterior distribution of λ (given that one HLA match was found in the sample of 5000).

Section 9.12

9-V In Problem 8-D we encountered the Maxwell distribution, with p.d.f.

$$f(u \mid \theta) = \sqrt{\tfrac{2}{\pi}} \cdot \frac{u^2}{\theta^{3/2}} \exp\left(-\frac{u^2}{2\theta}\right), \ u > 0.$$

We found that it is a member of the exponential family. Show that the m.l.e. of θ is efficient, given $E(U^2) = 3\theta$. [The p.d.f. is that of a variable V which is the distance from the origin of a point (X, Y, Z), where each component is $\mathcal{N}(0, \theta)$; so V^2 is $X^2 + Y^2 + Z^2$, with mean value 3θ.]

Chapter 9: Solutions

9-A **Solution:**

(a) The mean and s.d. are $\bar{X} = .02667$ and $S = .01102$. The standard error is

$$\text{s.e.} = \frac{S}{\sqrt{n}} = \frac{.01102}{\sqrt{6}} = .0045.$$

(b) To satisfy s.e. $\leq .0020$, we set $.011/\sqrt{n} = .0020$ and solve for n:

$$n = (.011/.002)^2 \doteq 30.$$

9-B **Solution:**

The m.l.e. of the population proportion is the sample proportion, $20/72$. The standard error is

$$\sqrt{\frac{\hat{p}(1-\hat{p})}{n}} = \sqrt{\frac{\frac{20}{72} \cdot \frac{52}{72}}{72}} \doteq .053.$$

9-C **Solution:**

(a) With no information about p, use (3) of §9.2: $n \geq .25/.05^2 = 100$.

(b) Use (4) of §9.2 with $p_0 = .8$: $n \geq .8 \times .2 /.05^2 = 64$.

9-D **Solution:**

The sample mean is always an unbiased estimate of the population mean, since $E\bar{X} = \mu$. So the bias in $2\bar{X}$ as an estimator of $\theta = 2\mu$ is also 0. The m.s.e. is then the variance:

$$\text{var}\, 2\bar{X} = 4\,\text{var}\,\bar{X} = \frac{4\sigma^2}{n} = \frac{\theta^2}{3n}.$$

(We've used the fact that the variance of a uniform distribution over an interval of length l is $l^2/12$.)

9-E **Solution:**

The sample proportion $\hat{p}$ is asymptotically normal with mean p and variance $p(1-p)/n$ (see §8.8). Thus, for given $\epsilon > 0$,

$$P(|\hat{p} - p| > \epsilon) = 1 - P(p - \epsilon < \hat{p} < p + \epsilon)$$
$$\doteq 1 - \Phi\left(\frac{p+\epsilon-p}{\sqrt{p(1-p)/n}}\right) + \Phi\left(\frac{p-\epsilon-p}{\sqrt{p(1-p)/n}}\right)$$
$$\to 1 - \Phi(\infty) + \Phi(-\infty) = 0.$$

According to definition (1) in §9.3, this shows consistency of $\hat{p}$.

9-F   **Solution:**
The standard error of the mean is $60.7/\sqrt{74} = 7.056$. If we assume approximate normality of the sample mean, the confidence limits are $75.7 \pm 1.96 \times 7.056$, or 75.7 ± 13.8.

Since durations are positive, the size of S in relation to the mean suggests that the sample observations are skewed to the right. Nevertheless, for a sample of size 74, the sample mean should be very close to normal. (We note in passing that the mean duration has questionable usefulness. One is more apt to be interested in the proportion of storms that last more than an hour, more than two hours, and so on.)

9-G   **Solution:**
We assume random sampling (justifiable or not—without assuming something we can't draw any conclusions). The standard error of $\hat{p}$ is

$$\text{s.e.} = \sqrt{\frac{.57 \times .43}{264}} = .0305.$$

Then 95% confidence limits are $.57 \pm 1.96 \times .0305$, or $.57 \pm .060$. So it's not clear where the "8 percentage points" comes from. (Confidence limits at 99% confidence are about $.57 \pm .08$.)

9-H   **Solution:**
The standard error is $3.0/\sqrt{15} \doteq .7746$. The population s.d. is not given, so it must be estimated from the sample, and we turn to the t-distribution for the proper multiplier of the s.e. Table IIIa gives the 97.5 percentile of $t(14)$ as 2.14, so the limits are $20.3 \pm 2.14 \times .7746$. or 20.3 ± 1.66.

9-I   **Solution:**
The p.d.f. of $W = \sum X_i$, which is $\text{Gam}(n, \lambda)$, is

$$f_W(w \mid n, \lambda) = \frac{\lambda^n}{\Gamma(n)} w^{n-1} e^{-\lambda w}, \ w > 0.$$

The distribution of $Y = \lambda W$ is therefore

$$f_Y(y) = f_W\left(\frac{y}{\lambda} \mid n, \lambda\right) \cdot \frac{1}{\lambda} = \frac{\lambda^n}{\Gamma(n)} \left(\frac{y}{\lambda}\right)^{n-1} \cdot \frac{1}{\lambda} e^{-\lambda(y/\lambda)} = \frac{y^{n-1}}{\Gamma(n)} e^{-y}.$$

This is pivotal, being free of λ.

9-J   **Solution:**
The p.d.f. (except for a constant factor) is $(1 + t^2/\nu)^{-\nu/2} \times (1 + t^2/\nu)^{-1/2}$. The second factor tends to 1 as ν becomes infinite. The first factor can be written, with $x = t^2/\nu$, and $\nu = t^2/x$ in the form

$$[(1 + x)^{1/x}]^{-t^2/2}.$$

The quantity in brackets converges to e as x goes to 0 (ν goes to infinity).

9-K **Solution:**

This is just a matter of substituting $\nu = 1$ in the formula for the p.d.f., which is shown as (4) in §9.6:

$$f_{T(1)}(t) \propto \frac{1}{1 + t^2}.$$

9-L **Solution:**

We learned in §5.2 that, for any c.d.f. $F(x)$, the c.d.f. of $F(X)$ is uniform on the interval $(0, 1)$. Thus, if $V = F_T(T \mid \theta)$, then $V \sim \mathcal{U}(0, 1)$, which does not involve θ—so V is pivotal.

9-M **Solution:**

For 90% confidence limits, say, we'd obtain the 5th and 95th percentiles of $\text{Gam}(n, 1)$: $\gamma_{.05}$ and $\gamma_{.95}$. Then

$$.90 = P(\gamma_{.05} < \lambda \sum X_i < \gamma_{.95}) = P\left(\frac{\gamma_{.05}}{\sum X_i} < \lambda < \frac{\gamma_{.95}}{\sum X_i}\right).$$

The statistics surrounding λ in the last expression are the desired limits.

9-N **Solution:**

(a) The estimate of δ is the difference in sample means, $\widehat{\delta} = 65.8 - 15.6$. The s.e. of $\widehat{\delta}$, the difference in sample means is the square root of the sum of the squares of the individual s.e.'s:

$$\text{s.e.}(\widehat{\delta}) = \sqrt{\frac{14.9^2}{28} + \frac{3.8^2}{28}} \doteq 2.9.$$

(b) $\widehat{\delta}$ is approximately normal, so for 95% confidence, we use $1.96 \times$ s.e.:

$$\widehat{\delta} \pm 1.96 \times \text{s.e.}(\widehat{\delta}) = 50.2 \pm 1.96 \times 2.9 = 50.2 \pm 5.7.$$

9-O **Solution:**

We apply (5) of §9.8. The s.e. is

$$\sqrt{\frac{\widehat{p}_1(1 - \widehat{p}_1)}{n_1} + \frac{\widehat{p}_2(1 - \widehat{p}_2)}{n_2}} = \sqrt{\frac{.65 \times .35}{1200} + \frac{.48 \times .52}{950}} = .021.$$

So the desired limits are $.17 \pm 1.96 \times .021$, or about .13 to .21.

9-P **Solution:**

The bias in Y is $b_Y(\sigma^2) = EY - \sigma^2 = K(n-1)\sigma^2 - \sigma^2$. Its variance is

$$\text{var } Y = K^2(n-1)^2 \text{var } S^2 = 2K^2(n-1)\sigma^4.$$

The m.s.e. is

$$\text{var } Y + b_Y^2(\sigma^2) = 2K^2(n-1)\sigma^4 + [K(n-1) - 1]^2 \sigma^4$$
$$= \sigma^4[(n^2 - 1)K^2 - 2(n-1)K + 1].$$

This quadratic function of K has its minimum value when $K = 1/(n+1)$. The estimator with this value of K is $\frac{n-1}{n+1}S^2$; the bias is $-2\sigma^2/(n+1)$.

9-Q Solution:

The sample s.d. is 9.27. A dot diagram (make one!) does not reveal any obvious nonnormality, so we assume normality and follow the pattern of Example 9.9b. With 13 d.f. $(n-1)$, we find the 5th and 95th percentiles of $\text{chi}^2(13)$ in Table Va to be 5.89 and 22.4. The 90% limits for σ^2 are thus $13 \times 9.27^2/22.4$ and $13 \times 9.27^2/5.89$; taking square roots gives us the limits for σ: $7.06 < \sigma < 13.77$.

9-R Solution:

(a) The likelihood function is $\alpha^n \prod x_i^{\alpha-1}$, with logarithm

$$\log L(\alpha) = n\log\alpha + \sum(\alpha-1)\log x_i.$$

Differentiating and setting the derivative equal to zero yields the m.l.e.:

$$\widehat{\alpha} = \frac{-n}{\sum\log x_i}.$$

(b) The population mean is $\mu = \frac{\alpha}{\alpha+1}$. Solving for α we get $\alpha = \frac{\mu}{1-\mu}$. Replacing μ by $\bar{X}$ gives us the estimator $\frac{\bar{X}}{1-\bar{X}}$.

9-S Solution:

(a) The likelihood function has its maximum at the largest observation. This is the m.l.e.

(b) The population mean is $\mu = \theta/2$. Solving for θ we get $\theta = 2\mu$. With $\bar{X}$ in place of μ, we see that the estimator is $2\bar{X}$.

9-T Solution:

The likelihood function is $L(\theta) = \theta^n e^{-\theta\sum x^2/2}$, $\theta > 0$, Multiplying g and L we obtain the posterior:

$$h(\theta\mid\mathbf{X}) = \theta^{n+\alpha-1}e^{-\theta(\sum X^2/2+\lambda)}.$$

The estimator we seek is the mean of this posterior distribution, which is $\text{Gam}(n+\alpha, \frac{1}{2}\sum X_i^2 + \lambda)$; the mean is the ratio of the parameters:

$$\frac{n+\alpha}{\sum X_i^2 + \lambda}.$$

9-U   Solution:

The distribution of the number of compatible HLA's in a random sample of 5000 is $\text{Bin}(5000, p)$, but we'll approximate it with $\text{Poi}(\lambda)$, where $\lambda = 5000p$. Then, given 1 in 5000 as the data,

$$L(\lambda) = P(\text{one HLA} \mid \lambda) = \lambda e^{-\lambda} = h(\lambda \mid \text{data}),$$

since the prior is uniform. The posterior c.d.f. is thus

$$H(\lambda \mid \text{data}) = \int_0^\lambda u e^{-u} du = 1 - e^{-\lambda}(1 + \lambda).$$

We could set this equal to .025 and .075 and solve for the desired percentiles, but these require numerical solutions. You can check that the percentiles are .243 and 5.57. [We found these by noticing that the second term on the right is the sum of the Poisson probabilities for 0 and 1 when $m = \lambda$; and in the Poisson table, we found .974 for $m = .25$ and .982 for $m = .20$. Interpolating gave us the 2.5 percentile as .243. A similar interpolation produced the 5.57.]

9-V   Solution:

The likelihood function is

$$L(\theta) = \theta^{-3n/2} \exp\left(-\sum \frac{U_i^2}{2\theta}\right),$$

with logarithm

$$\log L(\theta) = -\frac{3n}{2} \log \theta - \sum \frac{U_i^2}{2\theta}.$$

The derivative,

$$\frac{\partial}{\partial \theta} \log L(\theta) = -\frac{3n}{2\theta} + \frac{1}{2\theta^2} \sum U_i^2,$$

vanishes at the m.l.e.: $\hat{\theta} = \frac{1}{3n} \sum U_i^2$. This is unbiased: $E\hat{\theta} = \frac{1}{3} E(U^2) = \theta$.

And since it is sufficient (a 1-1 function of the sample sum of squares, which is clearly sufficient, in view of the likelihood), it is efficient.

CHAPTER 10: Additional Problems

Sections 10.1-10.3

10-A Test the hypothesis that the mean number of arrivals in a Poisson arrival process is $\lambda = 1$ against the alternative $\lambda > 1$, given these numbers of arrivals in 10 successive unit time intervals: 2, 0, 1, 4, 0, 2, 2, 6, 3, 0.

10-B Consider testing H_0: $\theta = 0$ against H_A: $\theta > 0$, where θ is the median of the Cauchy distribution with p.d.f.
$$f(x \mid \theta) = \frac{1/\pi}{1 + (x - \theta)^2}.$$
Find the *P*-value in each case:
(a) We take one observation and obtain $X = 4$.
(b) We take 100 independent observations and find $\bar{X} = 4$.
(c) We take 3 independent observations and find the median: $\tilde{X} = 4$.

10-C Called to jury duty, one of the authors noticed that the panel of about 80 prospective jurors was all white. The population from which the panel was presumably chosen includes about 10% blacks. Is there evidence that there is bias in the selection process? (Assume random sampling. The selections are supposed to be made at random from various lists, but one who is so chosen may be excused from duty for valid reasons.)

10-D A college administrator takes a random sample of 100 from more than 10,000 applicants. The sample average ACT score is 21.6, and the standard deviation is $S = .49$. The average score nationwide in that particular year was 20.2. Is the mean score of all applicants to that college higher than the national average?

10-E Suppose $X \sim \mathcal{N}(0, \sigma^2)$, and we want to test H_0: $\sigma^2 = 1$ against the alternative H_A: $\sigma^2 \neq 1$, using a random sample of size $n = 80$. What do you conclude if it turns out that $\bar{X} = -.0320$ and $S = 1.07$?

Section 10.4

10-F Ten children were treated with a therapy involving the drug ethosuximide to see if it would increase IQ scores.[1] Increases in verbal IQ scores were observed as follows: 16, 7, -5, 24, 19, 11, 6, 2, 10, 30. Do these data indicate an increase in mean score for treated children?

---

[1]W. L. Smith, "Facilitating verbal-symbolic functions in children with learning problems . . .," in W. Smith (Ed.) *Drugs and Cerebral Function* (Thomas, 1970), page 125.

10-G Whiling away the hours in a hospital stay, we wanted to know if the sand
in a "3-minute" egg timer (that came in a game of "Boggle") actually
took 180 seconds to run out. We timed it twice, obtaining this sample:
208, 198. Test the hypothesis that the mean time is 180.

Section 10.5

10-H To test the hypothesis $\mu = 0$ in a symmetric, continuous population, we
take a random sample of size 9. The sequence of 9 observations defines a
sequence of signs (+ or −), and we can calculate the signed-rank statistic
R_- for each sample sequence.
(a) How many sign sequences are possible?
(b) Find the one-sided P-value for $R_- = 5$, by listing and counting sign
sequences for which $R_- \leq 5$, and so verify the entry in Table VI.
(c) Find the (one-sided) P-value for $R_- = 40$.

10-I In a study assessing the effectiveness of a drug intended to reduce VPB's
(ventricular premature beats), the following decreases in VPB's were
recorded, about 20 minutes after administration of 2 mg/kg of the drug:[2]

$$1, 7, 17, 22, 5, 4, 5, 14, 9, 7, -4, 51.$$

Test the hypothesis that there is no treatment effect against the research
hypothesis that the drug is effective in reducing VCB's "on average."

Section 10.6

10-J Suppose our prior for μ is flat (an improper prior, but proceed anyway).
Find the probability of the hypothesis $\mu \leq 0$, if we observe $\bar{X} = .5$ in a
random sample of size 100 from $\mathcal{N}(\mu, 4)$.

10-K A treatment for respiratory failure in newborn infants termed ECMO
(extracorporeal membrane oxygenation) was the object of a randomized
study in 1985.[3] The result was that 11 patients given ECMO survived,
and the only patient in the control group (standard care) died. (The
scheme for assignment to therapy or control favored the therapy if it had
been more successful in preceding trials.) Let p_T and p_C denote the
probabilities of survival under the treatment and control, respectively.
Assume a uniform distribution in the unit square for (p_T, p_C), and find
the probability of the alternative $p_T > p_C$.

---

[2]Reproduced in D. A. Berry, "Logarithmic transformations in ANOVA," *Biometrics*
43 (1987), 439-456.

[3]L. P. Novak, "Working capacity, body composition, and anthropometry of Olympic
female athletes," *J. Sports Med. and Physical Fitness* **17**, 275-283.

Chapter 10: Solutions

10-A Solution:
The sample sum, the total number of arrivals in the 10 intervals of time is sufficient for λ [Problem 8-9(c)]. Its distribution is $\text{Poi}(10\lambda)$, and we find the P-value in Table IV, corresponding to the observed $\sum X_i = 20$:

$$P = P(\sum X_i \geq 20) = 1 - P(\sum X_i \leq 19) = 1 - .997 = .003.$$

(The .997 is the entry in the column for $m = 10$ opposite $c = 19$.) The result is termed "highly statistically significant."

10-B Solution:
(a) The P-value is the probability to the right of $x = 4$ when $\theta = 0$, which is 1 minus the c.d.f. at 4:

$$P(X > 4 \mid \theta = 0) = 1 - F(4 \mid 0).$$

The c.d.f. is $\frac{1}{2} + \frac{1}{\pi}\text{Arctan}(x - \theta)$; and $\text{Arctan}(4 - 0) = 1.326$, so

$$P = 1 - (.5 + \frac{1.326}{\pi}) = 1 - .922 = .078.$$

(b) The distribution of $\bar{X}$ is the same as the distribution of X [see Problem 8-Q). So again $P = .078$

(c) When $n = 3$, the median is the second smallest observation, $X_{(2)}$. We do have a formula for the p.d.f. of the median [(4) in §8.6], but the tail-area we want is $1 - F_{X_{(2)}}(4)$, so we'll get the c.d.f.:

$$F_{X_{(2)}}(y) = P(2 \text{ or } 3 \ X\text{'s} \leq y).$$

The number of X's $\leq y$ is $\text{Bin}(3, F_X(y))$, so

$$F_{X_{(2)}}(y) = F_X^3(y) + 3F_X^2(y)[1 - F_X(y)],$$

where

$$F_X(y) = \frac{1}{2} + \frac{1}{\pi}\text{Arctan}\, y, \text{ and } F_X(4) = .5 + \frac{1}{\pi}\text{Arctan}\, 4 = .922.$$

$$P = P(X_{(2)} \leq 4) = 1 - F_{X_{(2)}}(4) = 1 - (.922^3 + 3 \times .922^2 \times .078) = .0173.$$

10-C Solution:
Interpreting this result is problematical. Every sample has some feature that will seem unusual. Merely finding an unusual feature is not evidence against random sampling. However, suppose, prior to assembling with the rest of the panel, we had in mind to observe the number of blacks selected, and to use that number in a test of the hypothesis that the population from which jurors were selected was 10% black, against the alternative

that it was less than 10% black. For the result we found (0 blacks in a sample of size 80), the P-value is

$$P = P(0 \text{ blacks} \mid p = .1) = .9^{80} \doteq .0002.$$

10-D   Solution:

We can't answer exactly the question as stated. What we can do is to test the hypothesis that the mean of those applying is not different from the national average: $\mu = 20.2$, against the alternative $\mu > 20.2$. The appropriate Z-score is

$$Z = \frac{21.6 - 20.2}{4.9/\sqrt{100}} = \frac{1.4}{.49} \doteq 2.9.$$

The tail-area beyond 2.9 is $P \doteq .002$.

10-E   Solution:

The m.l.e. of σ^2 is $\frac{1}{80} \sum X_i^2 = \frac{79}{80} \times 1.07^2 + .0320^2 = 1.1316$. The mean of this estimator is $E(X^2) = \sigma^2$, and the variance is $\frac{\text{var } X^2}{n} = \frac{2\sigma^2}{n}$. So, by the central limit theorem, the asymptotic distribution of this estimator is $\mathcal{N}(\sigma^2, 2\sigma^2/n)$, or (under H_0), $\mathcal{N}(1, 2/80)$. Thus, we can use a Z-statistic:

$$Z = \frac{1.1316 - 1}{\sqrt{2/80}} = .832.$$

The observed value of the test statistic is not quite as far from $\sigma = 1$ as is typical when H_0 is true—not far enough to cast doubt on H_0.

10-F   Solution:

We test H_0: $\mu = 0$ (where μ is the mean increase in question) against the alternative (the research hypothesis) H_A: $\mu > 0$. The sample mean and s.d. are $\bar{X} = 12$, and $S = 10.48$. Not knowing the population variance, we have to estimate σ using S; so we calculate the t-score:

$$T = \frac{12.0 - 0}{10.48/\sqrt{10}} = 3.62.$$

In view of the alternative, large values of T are extreme, and we find the P-value in Table IIIb, under the heading 9 d.f. and opposite 3.62 (or as close as we can get to 3.62): $P \doteq .003$, fairly strong evidence against the null hypothesis.

(Even if convinced by this evidence that the mean score has increased, we can't jump to the conclusion that the drug therapy was responsible. In taking tests, subjects learn; they tend to do better on a second attempt. Any effect the therapy may have is confounded with the learning, which makes the study worthless for judging the effect of the therapy. One way to design a proper study, taking learning into account, is to have a second group of children who take the IQ test twice, but are given a placebo or no treatment between tests.)

10-G Solution:
The sample mean is 203, and the s.d. is $S = \sqrt{50} = .707$. With these,
$$T = \frac{203 - 180}{\sqrt{50}/\sqrt{2}} = 4.6.$$
Table IIIa shows this to be between the 90th and 95th percentile of $t(1)$, but Table IIIb doesn't list entries for 1 d.f. But we can get an exact P-value by using the fact that the distribution of $t(1)$ is Cauchy [see (5) of §9.6]:
$$P(T > 4.6) = \int_{4.6}^{\infty} \frac{1/\pi}{1 + x^2}\, dx = \tfrac{1}{\pi}(\text{Arctan } \infty - \text{Arctan } 4.6) = .068.$$
This is 1-sided; if you feel it should be doubled, feel free.

10-H Solution:
(a) Since each sequence element can be $+$ or $-$, there are $2^9 = 512$ possible sign sequences.

(b) Here are sequences with values of R_- up to 5:

Sequence	R_-
$+ + + + + + + + +$	0
$- + + + + + + + +$	1
$+ - + + + + + + +$	2
$+ + - + + + + + +$	3
$- - + + + + + + +$	3
$- + - + + + + + +$	4
$+ + + - + + + + +$	4
$+ - - + + + + + +$	5
$- + + - + + + + +$	5
$+ + + + - + + + +$	5

Each sequence has probability 1/512, so $P(R_- \le 5) = 10/512 \doteq .0195$.

(c) The sum of the integers 1 to 9 is 45, so the point symmetric to 40 in the distribution of R_- is $45 - 40 = 5$. The right-tail probability for 40 is the same as the left-tail probability for 5: $P = P(R_- \ge 40) = 10/512$.

10-I Solution:
The usual summary statistics are $\bar{X} = 11.5$, $S = 14.286$. With these, we calculate
$$T = \frac{11.5}{14.286/\sqrt{12}} \doteq 2.79.$$
Table IIIb (11 d.f.) gives a P-value of about .009. But, is the t-test really appropriate? If we omit the most successful case (51), then $\bar{X} = 7.91$, $S = 7.37$, and $T = 3.56$—with P less than .003, *more* significant than with the largest decrease *in* the sample. A dot diagram (make one) will suggest

the reason: The population is apparently not close to normal—a t-test is not a good idea.

We could test the hypothesis that the median decrease is 0 using the sign test: 11 +'s and 1 − yields $P = .0029$ (in Table Ib, $n = 12$, $k = 11$). If the population symmetry is in question for a t-test, it may be also questioned for a signed rank test (which yields $P = .001$). Another approach is to transform the data so that normality is more plausible, and then use a t-test on the transformed data.

10-J   Solution:
The posterior is proportional to the likelihood (if the prior is constant), which we get from Example 8.2c as (2):

$$h(\mu \mid \text{data}) \propto e^{-\frac{n}{8}(\mu - \bar{X})^2},$$

which is the p.d.f. of $N(\bar{X}, 4/100)$. And then, since $\bar{X} = .5$,

$$P(\mu \leq 0) = \Phi\left(\frac{0 - .5}{.2}\right) = .0062.$$

10-K   Solution:
If we assume independence of the control and treatment samples, the joint likelihood function is the product of the individual likelihoods:

$$L(p_T, p_C) \propto p_T^{11}(1 - p_T)^0 \times p_C^0(1 - p_C)^1.$$

Given that the prior is uniform, the posterior is proportional to this:

$$h(p_T, p_C) = \frac{p_T^{11}(1 - p_C)^1}{\displaystyle\int_0^1 \int_0^1 p_T^{11}(1 - p_C)^1 \, dp_T \, dp_C} = 24 p_T^{11}(1 - p_T)^1.$$

Then

$$P(p_T > p_C) = 24 \int_0^1 \int_0^x x^{11}(1 - y) \, dy \, dx = 24 \int_0^1 x^{11}(x - \tfrac{1}{2}x^2) \, dx = \frac{90}{91}.$$

CHAPTER 11: Additional Problems

Sections 11.1-11.3

11-A The sample space of a test statistic X has 5 values: $\{a, b, c, d, e\}$.
Consider testing f_0 vs. f_1, probability functions for X, defined thus:

x	$f_0(x)$	$f_1(x)$
a	.1	.3
b	.1	0
c	.2	.2
d	.3	.4
e	.3	.1

(a) Find α and β for the rejection region $\{b, c\}$.
(b) Find α and β for the rejection region $\{d\}$.
(c) How many distinct rejection regions (good and bad) are possible?
(d) Give a rejection region which has $\alpha = 0$, and one with $\beta = 0$.

11-B Consider the rejection region $\bar{X} > 2$, where $\bar{X}$ is the mean of a random
sample of size 64 for testing H_0: $\mu = 1$, given $\sigma = 4$.
(a) Find α, the size of the type I error.
(b) Find the probability of rejecting H_0 when $\mu = 3$.

11-C In the setting of the preceding problem, suppose we decide to reject $\mu = 1$
if a 90% confidence interval calculated from the sample of size 64 does
not include the value 1.
(a) Express this rule as an inequality for $\bar{X}$.
(b) Find α for this rule.

Sections 11.2-11.3

11-D Find the power function of the decision rule given in Problem 11-B.

11-E Find the power function of the decision rule given in Problem 11-C.

11-F Again in the setting of Problem 11-B, suppose we use the rejection region
$\bar{X} > K$. Find the sample size n and the value of K so that $\alpha = .01$ and
the probability of accepting $\mu = 1$ when in fact $\mu = 2$ is .01.

Section 11.4

11-G An acceptance sampling scheme is to have probability .05 of rejecting a lot in which the lot fraction defective is $p = .04$, and a probability .05 of accepting a lot in which $p = .20$. Find an appropriate sample size and acceptance number.

Sections 11.5–11.7

11-H Referring to the setting of Problem 11-A, find the most powerful rejection regions, and give the α and β for each.

11-I Referring to the preceding problem, find a most powerful rejection region with size $\alpha = .2$.

11-J Consider a single observation for testing H_0: $X \sim N(0, 1)$ against the alternative that X has a Cauchy distribution with median 0:

$$f_1(x) = [\pi(1 + x^2)]^{-1}.$$

Find the most powerful rejection regions.

11-K Find the UMP tests for $\sigma = 1$ against $\sigma > 1$, based on a random sample of size n from $N(0, \sigma^2)$.

Section 11.8

11-L Construct a likelihood ratio test for $\theta = \theta_0$ vs. $\theta \neq \theta_0$, based on a random sample of size n from $N(\mu, \theta)$.

11-M Apply the likelihood ratio method to testing $\lambda = 2$ against $\lambda \neq 2$, given a random sample from Gam(2, λ). Determine what kinds of values of $\bar{X}$ are extreme, offering evidence against $\lambda = 2$.

Section 11.9

11-N Consider again the setting of Problem 11-A, where the sample space of X has 5 values: $\{a, b, c, d, e\}$, and we test f_0 vs. f_1, defined thus:

x	$f_0(x)$	$f_1(x)$
a	.1	.3
b	.1	0
c	.2	.2
d	.3	.4
e	.3	.1

(a) Suppose a prior distribution assigns probabilities .7 and .3, to f_0 and f_1, respectively. Assume losses of 4 if we reject f_0 when it is true, and 2 if we accept f_0 when it is false (and 0 if we make the right decision). With no data, what is the correct action (the one that minimizes expected loss)?

(b) Find the posterior probability of each model, given that we observe $X = a$.

(c) With the losses as defined in (a), find the Bayes action after we have observed $X = a$.

Chapter 11: Solutions

Section 11.1

11-A Solution:
(a) The type I error size is the probability of the rejection region under H_0: $.1 + .2 = .3$. The type II error size is the probability of the complement of the critical region under H_1: $P_1(a, d, \text{ or } e) = .8$.

(b) $\alpha = P_0(d) = .3$, and $\beta = 1 - P_1(d) = 1 - .4 = .6$.

(c) For each sample point, we have two choices—either put it in the rejection region or don't: 2^5 possibilities.

(d) The rejection region $\emptyset$ always has $\alpha = 0$, and the region Ω always has $\beta = P(\Omega^c) = 0$. [In this case, there is another acceptance region with probability 0: $\{b\}$.]

11-B Solution:
(a) By definition, $\alpha = P(\bar{X} > 2 \mid \mu = 1)$. Since $E\bar{X} = \mu = 1$, and var $\bar{X} = 4/8 = .5$ (and if we assume 64 is large enough for $\bar{X}$ to be close to normal),
$$\alpha \doteq 1 - \Phi\left(\frac{2-1}{.5}\right) = \Phi(-2) = .0228.$$

(b) $P(\bar{X} > 2 \mid \mu = 3) = 1 - \Phi\left(\frac{2-3}{.5}\right) = .9772.$

11-C Solution:
(a) The 90% confidence limits are $\bar{X} \pm 1.645 \times .5$, so the rule is to reject H_0 unless $\bar{X} - .8225 < 1 < \bar{X} + .8225$.

(b) $\alpha = P(\text{reject } H_0 | H_0) = 1 - P(\bar{X} - .8225 < 1 < \bar{X} + .8225 \mid H_0)$
$$= 1 - P(|\bar{X} - 1| < .8225 \mid \mu = 1)$$
$$= 1 - P\left(\frac{|\bar{X} - 1|}{.5} < 1.645 \mid \mu = 1\right) = 1 - .90 = .10.$$

[This works out this way because when $\mu = 1$, the ratio $(\bar{X} - 1)/.5$ is standard normal.]

That is, if the probability is 90% that the interval will include the true value of μ, the probability is 10% that it will not include the value 1 when the true value is 1. (Can you generalize?)

11-D Solution:

The calculation in 11-B(b) is a special case, finding the power at $\mu = 3$. So the calculation in general will be just like what we did there, except with μ left as μ:

$$\pi(\mu) = P(\bar{X} > 2 \mid \mu) = 1 - \Phi\left(\frac{2 - \mu}{.5}\right) = \Phi(2\mu - 4).$$

With this power function, we get α as $\pi(1) = \Phi(-2)$, as before.

11-E Solution:

The calculation of the power function proceeds much like what we did in finding α, except that we have an arbitrary μ in place of $\mu = 1$:

$$\pi(\mu) = 1 - P\left(\frac{|\bar{X} - 1|}{.5} < 1.645 \mid \mu\right) = 1 - P(1 - .8225 < \bar{X} < 1 + .8225 \mid \mu)$$

$$= 1 - \left\{\Phi\left(\frac{1.8225 - \mu}{.5}\right) - \Phi\left(\frac{.1775 - \mu}{.5}\right)\right\}.$$

11-F Solution:

For these conditions, we want $\pi(1) = .01$, and $\pi(2) = .99$, where

$$\pi(\mu) = P(\bar{X} > K \mid \mu) = 1 - \Phi\left(\frac{K - \mu}{4/\sqrt{n}}\right).$$

These two conditions tell us that

$$\frac{K - 1}{4/\sqrt{n}} = 2.326, \text{ and } \frac{K - 2}{4/\sqrt{n}} = -2.326.$$

Subtracting the second of these equations from the first gives $\sqrt{n}/4 = 4.652$, or $\sqrt{n} = 18.61$, $n = 347$. Substituting in either equation gives $K = 1.5$, which is not too surprising (half-way between 1 and 2), in view of the symmetry.

11-G Solution:

The stated requirements can be expressed as these equations:

$$\sum_0^c f(k \mid p = .04) = .95, \quad \sum_0^c f(k \mid p = .20) = .05,$$

where c is the acceptance number, and f is basically hypergeometric. But no lot size is specified, so we'll have to assume it is "large"—large enough that the distribution of the number of defectives is $\text{Bin}(n, p)$. And with a small value of p perhaps n will be large enough that a Poisson approximation will do.

Solving the equations for c and k is difficult, and with the requirement that these be integers, there may be no exact solution. Trial and error may take a while. We'll just show that $n = 36$ and $c = 3$ are approximate solutions. When $p = .04$, the mean is $np = 1.44$, and

$$\sum_0^3 f(k \mid .04) = e^{-1.44}\left(1 + 1.44 + \frac{1.44^2}{2} + \frac{1.44^3}{3!}\right) \doteq .94.$$

When $p = .20$, $np = 7.2$ and $npq = 5.76$, we use a normal approximation:

$$\sum_0^3 f(k \mid .20) \doteq \Phi\left(\frac{3.5 - 7.2}{\sqrt{5.76}}\right) = \Phi(-1.54) = .062.$$

11-H Solution:
We repeat the table of Problem 11-A, with an additional column of values of the likelihood ratio:

x	$f_0(x)$	$f_1(x)$	$\Lambda(x)$
a	.1	.3	1/3
b	.1	0	∞
c	.2	.2	1
d	.3	.4	3/4
e	.3	.1	3

The values of X arranged with increasing Λ are: a, d, c, e, b. So the Neyman-Pearson rejection regions are $\emptyset$, $\{a\}$, $\{a, d\}$, $\{a, d, c\}$, $\{a, d, c, e\}$, and the whole sample space. For α, we look in the table under f_0 for the probabilities of the rejection region; for β, we look under f_1 for the probabilities of the complement of the rejection region:

R	α_R	β_R
$\emptyset$	0	1
$\{a\}$	.1	.7
$\{a, d\}$	.4	.3
$\{a, d, c\}$	.6	.1
$\{a, d, c, e\}$	.9	0
Ω	1	0

11-I Solution:
None of the tests defined in Problem 11-H has $\alpha = .2$, so achieving this requires randomization. The region $\{a\}$ has $\alpha = .1$, but adding d to the region increases α by .3, which is too much. So we include d only part of the time, rejecting H_0 with probability γ if we observe $X = d$. The α for this randomized rule is

$$\alpha = P(\text{reject } H_0 \mid H_0) = P(X = a) + \gamma \times P(X = b) = .1 + .3\gamma.$$

This will be .02 if $\gamma = 1/3$. So the rule is: Reject H_0 if $X = a$, and reject H_0 with probability $1/3$ if $X = d$, but otherwise accept H_0.

11-J **Solution:**
The p.d.f. under H_0 is $f_0(x) = (2\pi)^{-1/2} e^{-x^2/2}$. So the likelihood ratio for a Neyman-Pearson test (which is most powerful) is

$$\Lambda < C' \cdot \frac{1 + X^2}{\exp(X^2/2)}, \quad \text{or} \quad \log(1 + X^2) - \tfrac{1}{2}X^2 < C.$$

This is small when X is large. (It is 1 when $X^2 = 0$, increases for a bit as X^2 increases, then falls off rapidly to 0.) So regions of the form $X^2 > K$ are most powerful.

11-K **Solution:**
If we take a particular σ in the alternative, the Neyman-Pearson region for testing the value 1 as H_0 against σ as H_1 is

$$\Lambda = \frac{e^{-\Sigma x^2/2}}{\frac{1}{\sigma^{n/2}} e^{-\Sigma x^2/(2\sigma^2)}} = \sigma^{n/2} \exp\left\{-\tfrac{1}{2}\left(1 - \tfrac{1}{\sigma^2}\right)\sum X_i^2\right\} < K.$$

Since $\sigma > 1$, the quantity in parentheses is positive, and the exponent is negative. Therefore, Λ will be smaller than the constant K if $\sum X_i^2$ is larger than a constant K'. And this is true for *each* σ in the alternative set, so such regions are UMP against $\sigma > 1$.

11-L **Solution:**
The likelihood function is

$$L(\mu, \theta) = \theta^{-n/2} \exp\left\{-\tfrac{1}{2\theta}\sum (X_i - \mu)^2\right\}.$$

The maximum over all parameter values is achieved (by definition thereof) the maximum likelihood estimators, $\bar{X}$ and $V = \frac{n-1}{n} S^2$. The maximum likelihood is thus

$$V^{-n/2} \exp\left\{-\tfrac{1}{2V}\sum (X_i - \bar{X})^2\right\} = V^{-n/2} \exp(-n/2).$$

For the numerator of the likelihood ratio, we need the maximum over μ, holding θ fixed at θ_0. With θ fixed, the only variation is in the exponent, and we maximize the exponential by minimizing $\sum (X_i - \mu)^2$—which we know is accomplished by setting μ equal to $\bar{X}$. Thus, the maximum of L under H_0 is

$$L(\bar{X}, \theta_0) = \theta_0^{-n/2} \exp\left\{-\tfrac{1}{2\theta_0}\sum (X_i - \bar{X})^2\right\} = \theta_0^{-n/2} \exp\left\{-nV/(2\theta_0)\right\}.$$

So the likelihood ratio is

$$\Lambda = \frac{L(\bar{X}, \theta_0)}{L(\bar{X}, V)} = \frac{\theta_0^{-n/2} \exp\left\{-nV/(2\theta_0)\right\}}{V^{-n/2} \exp(-n/2)} = \left\{\frac{\exp[V/\theta_0]}{e \cdot (V/\theta_0)}\right\}^{-n/2}.$$

If we let $W = V/\theta_0$, we see that the likelihood ratio is small when e^W/W is large. The function e^W/W is concave up with a minimum at $W = 1$; so it will be large when either W is very small, or W is very large. So values of V that are much bigger or much smaller than θ_0 are considered "extreme." For a given K, the region of values of V corresponding to $\Lambda < K$ is two-sided, but not symmetric about θ_0, nor are the tail-areas under H_0 equal—which is awkward, but that's the way it is.

11-M Solution:

The p.d.f. is $\lambda^2 x e^{-\lambda x}$ for $x > 0$, so the likelihood function is $\lambda^{2n} e^{-\lambda \Sigma x_i}$. This is maximized (over all $\lambda > 0$) at $\hat{\lambda} = 2/\bar{X}$. The generalized likelihood ratio is

$$\Lambda = \frac{L(2)}{L(\bar{X})} = \frac{2^{2n} e^{-2n\bar{X}}}{(2/\bar{X})^{2n} e^{-2n\bar{X}/\bar{X}}} = \bar{X}^{2n} e^{-2n(\bar{X}-1)}.$$

This function of $\bar{X}$ goes to 0 as $\bar{X}$ goes to 0 and to ∞; it will be small if $\bar{X}$ is either very small (close to 0) or very large.

11-N Solution:

(a) If we accept H_0, the expected loss is $0 \times .7 + 4 \times .3 = 1.2$. If we reject H_0, the expected loss is $2 \times .7 + 0 \times .3 = 1.4$. The action with the smaller expected loss is to accept H_0.

(b) The (unconditional) probability of $X = a$ is $.7 \times .1 + .3 \times .3 = .16$. Bayes' theorem tells us that the posterior probabilities are $h_0 = 7/16$ and $h_1 = 9/16$.

(c) Applying the posterior probabilities to the loss table, we find $0 \times 7 + 4 \times 9 = 36$, and $2 \times 7 + 0 \times 9 = 14$ as (except for the divisor 16) expected losses when accepting and rejecting H_0, respectively. The smaller is 14, so the Bayes action is now to reject H_0.

CHAPTER 12: Additional Problems

Sections 12.1-12.3

12-A A research study in dental therapeutics reported data on the use of "MFP Fluoride" (by Colgate) and a leading stannous fluoride toothpaste. The subjects were children, ages 9-13, in Texas, with meals and brushing at home. The data (number of new cavities over a 3-year period) are summarized as follows:

	$\bar{X}$	S	n
Colgate's MFP fluoride	4.60	2.40	172
Stannous flouride	4.83	2.29	193

Test the hypothesis that the population mean difference is 0. (Colgate sponsored the research, so it's a fair assumption that the alternative hypothesis—the research hypothesis—was one-sided.)

12-B A 1987 Gallup survey found "no evidence of growing public intolerance towards gays." The poll included 1,015 adults in "scientifically selected localities across the nation" during the period March 14-18. One question asked was this: "Do you think homosexual relations between consenting adults should or should not be legal?" In 1986, the percentage who said "should not" was 54; in 1987, it was 55. Despite the increase, Gallup does not see it as evidence of growing intolerance. Is this reasonable? (Assume that the number polled in 1986 was also 1,015, and that both samples were random samples.)

12-C The denominator of the Z-statistic for comparing proportions [(7) in §12.3] is the *standard error* of the difference in sample proportions. Assume equal sample sizes: $n_1 = n_2 = n$ and find the sample size that would ensure a standard error of no more than .03.

Sections 12.4-12.5

12-D Stress-related factors in particular career aspirations are the subject of a study that reported the following statistics on systolic blood pressure for people in two types of careers:[4]

---

[4]C. P. Benlow and J. C. Standby, "Sex differences in mathematical ability: Fact or artifact?" *Science* **210** (1980), 1262–1264.

	n	$\bar{X}$	S
Professional	9	120.4	9.6
Semi-skilled	14	110.1	8.4

Test the hypothesis that professional people have higher blood pressures on average than semi-skilled workers.

12-E A research study of female Olympic athletes reports that "runners and swimmers seemed to have signficantly broader pelvises" than gymnasts.[5] The data (bicristal diameter in cm) presented there are summarized as follows:

	$\bar{X}$	S	n
Runners and swimmers	28.0	.885	15
Gymnasts	25.9	.512	5

(a) Test the null hypothesis of no difference in mean diameter.
(b) Do you see any problem with the quoted conclusion?
(c) "Test" implies that the data are from certain populations. What would these be?
(d) Would you attribute any difference in population means to the way the bodies have been used (as the report seemed to suggest)?

12-F Example 9.8a gave summary statistics on amounts of water filtered with a new filtration device (in m^3/m^2 filter), using two methods of operation. The raw data were as follows:

Method 1:   309 365 221 172   81 277 286 248 243 119
            243 216 196 182 157 194 148 182 105   98

Method 2:   221 249 271 187   99 222 161 379 307 294
            305 325 280 308 311 258 203 400 354 296
            253 415 344 329 282 356 323 248   91

The summary statistics, repeated from Example 9.8a, are as follows:

Method	n	*Mean*	*s.d.*
1	20	202.0	74.35
2	29	278.3	79.03

Test the hypothesis of no difference between the two methods in mean amount filtered, using an appropriate *t*-statistic.

---

[5]"Contrasting patterns of blood pressure and related factors within a Maori and European population in New Zealand," *Social Science and Medicine* **23** (1986), 439–444.

Test the hypothesis of no difference between the two methods in mean amount filtered, using an appropriate t-statistic.

12-G An experiment with kittens compared pattern recognition of males and females. With the criterion 27 out of 30 correct responses to a visual stimulus, the number of trials to meet the criterion were recorded as follows:[6]

Males: 120, 130, 155, 150, 40, 106, 382, 76, 89
Females: 69, 117, 66, 391, 94, 103.

Test the hypothesis of no difference between males and females.

12-H Verify the entry for $c = 24$, $m = 6$, $n = 7$ in Table VII of Appendix 1. Use the fact that the possible sequences of 6 X's and 7 Y's are equally likely under the hypothesis of identical populations (and that we are dealing with independent, random samples).

12-I One can also apply the rank test with the data of Problem 12-E above. This requires putting the 49 observations in numerical order—no small task, but not impossible, by hand. We did it with computer software, and found $R_1 = 331$. Complete the rank-sum test for the hypothesis that the two population distributions are identical.

Section 12.6

12-J Plasma half-life data (in hours) of the drug verapamil, after administration of verapamil only (V) and after verapamil administered concurrently with another drug (V+), were recorded for eight subjects:

Subject	1	2	3	4	5	6	7	8
V	2.55	1.81	1.99	2.37	3.03	2.25	1.89	1.83
V+	3.15	2.07	3.22	2.67	2.90	2.47	1.31	2.68
Difference	.60	.26	1.23	.40	−.13	.22	−.58	.85

Test the hypothesis that, on average, the second drug has no effect on the plasma half-life of verapamil.

12-K Example c in the Prolog reported data from a study in ophthalmology. Corneal thickness (μ) was measured on each eye of eight subjects, each of whom had unilateral glaucoma (one eye affected).

---

[6]Dodwell, Wilkinson, and von Grunan, "Pattern recognition in kittens," *Perception* **12** (1983), 393–410.

Subject	1	2	3	4	5	6	7	8
High IOP eye	488	478	480	426	440	410	458	460
Lo IOP eye	484	478	492	444	436	398	464	476

Test the hypothesis of no difference in thickness of the cornea (on average) between glaucomatous and normal eyes. (Glaucoma is characterized by high intraocular pressure (IOP):

Sections 12.7-12.8

12-L Find the 5th percentile of $F(4, 10)$.

12-M Problem 12-16 gives data on platelet counts of 10 normal individuals and 14 individuals with a recent thrombosis. The sample standard deviations are 53.4 and 139.1, respectively. Assuming normal populations, test the hypothesis of equality of population variances.

Chapter 12: Solutions

Section 12.1

12-A Solution:

The samples are quite large, and a Z-test is in order:

$$Z = \frac{.23}{\sqrt{\frac{2.4^2}{172} + \frac{2.29^2}{193}}} \doteq .91.$$

The observed difference is very close to what one would expect under the hypothesis of no difference in population means. There's no reason to doubt H_0 on the basis of these data.

12-B Solution:

We test $p_1 = p_2 = p$ vs. $p_1 \neq p_2$. Under the null hypothesis, we need an estimate for p: $\hat{p} = (548 + 558)/2030 = .5448$. This is the ratio of the total number of "successes" in both samples divided by the size of the combined sample. The standard error is then

$$\sqrt{\hat{p}(1 - \hat{p})\left(\frac{1}{n_1} + \frac{1}{n_2}\right)} = \sqrt{.5448 \times .4552 \times \left(\frac{1}{1015} + \frac{1}{1015}\right)} = .0221.$$

The Z-score for the test is then

$$Z = \frac{10/1015}{.0221} = .446.$$

Being less than half of the deviation from 0 that one would typically find when H_0 is true (namely, .0221), the observed difference in sample is easily explained as sampling variability—there's no evidence here that the population proportions are different, and the stated conclusion is justified.

12-C Solution:

With the assumption of equal sample sizes and equal p's, the standard deviation of the estimator $\hat{p}_1 - \hat{p}_2$ is

$$\text{s.d.}(\hat{p}_1 - \hat{p}_2) = \sqrt{p(1 - p)(1/n + 1/n)} .$$

The worst case is $p = .5$, so we choose n so that the s.d. is less than .03 when we replace p by .5:

$$\sqrt{.5 \times .5 \times 2/n} \leq .03.$$

Solving for n, we find that this holds when $n \geq 556$. And this n is more than large enough when $\hat{p}$ is not .5.

12-D **Solution:**
With population variances unknown, and small sample sizes, we use a t-statistic. We also assume independent random samples and (at least under H_0) equal population variances. The pooled variance [(3) of §12.4] is

$$S_p^2 = \frac{1}{21}[8 \times 9.6^2 + 13 \times 8.4^2) = 78.79.$$

With this in (4), we have

$$T = \frac{120.4 - 110.1}{\sqrt{78.79(1/9 + 1/14)}} = \frac{10.3}{3.792} \doteq 2.72.$$

The 1-sided P, from Table IIIb, is .007.

12-E **Solution:**
(a) With small samples and unknown variance, we use a 2-sample t-test. For this, we need the pooled variance:

$$S_p^2 = \frac{14 \times .885^2 + 4 \times .512^2}{18} = .817^2.$$

The test statistic is then

$$T = \frac{28.0 - 25.9}{.817\sqrt{1/15 + 1/5}} = 4.98 \quad (18 \text{ d.f.}).$$

The P-value (Table IIIb) is .000 (to 3 decimal places).

(b) The way the conclusion is stated sounds as though what has been observed has some practical significance.

(c) The only populations would be those of all Olympic swimmers and runners, and all Olympic gymnasts—if each year's crops could be thought of as random samples.

(d) If there is a difference (as the evidence strongly suggests), it is not necessarily that the different forms of activity causes it. A more likely explanation is that those with different builds tend to take up different sports.

12-F **Solution:**
A rough dot plot of the data gives no reason to question the assumption of at least near normality, so we proceed with a 2-sample t-test. Since d.f. $= 20 + 29 - 2 = 47$, we're going to end up in the region of the t-table, where we use a normal approximation. This suggests we could use the Z-statistic (4) in §12.2. We'll do it both ways. First the t-test: The pooled variance is

$$S_o^2 = \frac{1}{47}(19 \times 74.35^2 + 28 \times 79.03^2) = 77.17^2.$$

Then

$$T = \frac{278.3 - 202.0}{77.17\sqrt{1/20 + 1/29}} = 3.40.$$

From Table IIIa we can see only that $P < .001$. The Z-statistic is

$$Z = \frac{278.3 - 202.1}{\sqrt{\frac{74.35^2}{20} + \frac{79.03^2}{29}}} = 3.44.$$

In Table IIa we see that P is between .0002 and .0003.

12-G Solution:

There are some obvious outliers, one in each sample, suggesting that the population distributions are not close to normal. The rank-sum test would seem the more appropriate. The combined ordered sample is

40 **66 69** 76 89 **94 103** 106 **117** 120 130 150 155 382 **391**.

To use Table VII, we'll calculate the rank-sum for the smaller sample, the females; these observations are shown above in boldface type. The ranks (counting from left to right) are 2, 3, 6, 7, 9, 15, with sum 42. Looking in Table VII ($m = 6$, $n = 9$), we find that 42 is not in the extreme left tail of the distribution. But since the mean (under H_0) is $6 \times 16/2 = 48$, the value 42 is not in the right tail either. All we can say from the table is that $P > .164$ (the last entry shown, for $R = 39$). If there is a population difference, these samples don't show it.

12-H Solution:

A pattern is defined by the rank position of the X's, and we'll list the ranks of the X's in the most extreme cases, with the corresponding values of the rank sum:

Sequence	R
1 2 3 4 5 6	21
1 2 3 4 5 7	22
1 2 3 4 6 7	23
1 2 3 4 5 8	23
1 2 3 5 6 7	24
1 2 3 4 5 9	24
1 2 3 4 6 8	24

Thus, there are 7 patterns with rank-sum of 24 or less, out of a possible $\binom{13}{6} = 1716$, so $P = 7/1716 \doteq .0041$. The table starts at 24 even though the smallest possible value is 21, because the probabilities are so small when $R < 24$. The entry given for 24 is .004.

12-I Solution:

The sample sizes take us beyond Table VII, so we use the approximate normality of R for large samples, and the Z-statistic given by (2) of §12.5. For this we need the mean: $20 \times (20 + 29 + 1)/2 = 500$, and the variance, $20 \times 29 \times 50/12 = 7250/3$. Then

$$Z = \frac{331 - 500}{\sqrt{7250/3}} \doteq -3.44,$$

with $P \doteq .0006$. (A continuity correction would be in order, but does not appreciably change the result: $Z = 3.43$ with this correction, which would add .5 to 331 in the Z-score.)

12-J Solution:

A dot diagram of the differences shows no obvious nonnormality, so we can use either a t-test or a signed-rank test. For the t-test, we need the mean difference $\bar{D} = .356$, and the s.d. of the differences, $S_D = .562$. Substituting these in the *one*-sample t-statistic [(1) of §10.4], we find

$$T = \frac{.356}{.562/\sqrt{8}} \doteq 1.79.$$

The P-value is found in Table IIIb to be .057.

A signed-rank test can also be used: Arrange the differences in order of magnitude:

$$-.13, \quad .22, \quad .26, \quad .40, \quad -.58, \quad .60, \quad .85, \quad 1.23.$$

The negative differences have ranks 1 and 5, so $R\_ = 1 + 5 = 6$. In Table VI ($n = 8$, $c = 6$) gives $P = .055$—very close to the P-value we got using the t-test. But observe that the sign test is not very effective: 2 $-$'s out of 8 is not all that surprising [when $P(-) = \frac{1}{2}$], with $P \doteq .14$.

12-K Solution:

The data are paired—not two independent samples. So we work with the differences calculated for each pair: $-4, 0, 12, 18, -4, -12, 6, 16$. The mean and s.d. are $\bar{D} = 4$, $S = 10.74$, so

$$T = \frac{4 - 0}{10.74/\sqrt{8}} = 1.05.$$

This is just about the deviation of $\bar{D}$ from 0 that one expects under H_0, so it does not put H_0 in doubt. The signed-rank statistic could also be used: Among the 7 nonzero observations, three are negative; their ranks (when all 7 are ordered by magnitude: $-4, -4, 6, 12, -12, 16, 18$) are 1, 2, and 4.5 (the average of the number 4 and 5 spots occupied by the two 12's). So $R\_ = 7.5$, with a P-value of about .17 (interpolating in Table VI between $c = 7$ and $c = 8$ under $n = 7$).

12-L   Solution:

The 5th percentile of $F(4, 10)$ is the reciprocal of the 95th percentile of $F(10, 4)$. The latter is given in Table VIIIb as 5.96, so the desired percentile is $1/5.96 = .168$.

12-M   Solution:

The ratio of sample variances is $(139.1/53.4)^2 \doteq 6.8$. Under H_0 the ratio of sample variances is $F(13, 9)$. These degrees of freedom are beyond the range of Table VIIIa, but Table VIIIb gives the 1% tail-probability values. It gives the 99th percentile of $F(12, 9)$ as 5.11, and of $F(15, 9)$ as 4.96. These are both less than 6.8, so we conclude that $P < .01$.

CHAPTER 13: Additional Problems

Sections 13.1-13.3

13-A   At one time, plain M&M candies came in just five colors: brown, yellow, orange, tan, and green, mixed in the proportions 4:2:2:1:1 by the manufacturer. We counted the colors in a 1-lb bag taken from the store shelf and found 234 brown, 101 yellow, 91 orange, 50 tan, and 42 green M&M's. Assuming the bag to contain a random selection, test the hypothesis that the population proportions are those reported by the manufacturer.

13-B   Use the birthweight from Problem 7-B for a chi-square test of the hypothesis that the population is normal. The stem-leaf diagram of the data is repeated here. [The class intervals will have to divide the entire x-axis, so you'll have to start and finish with open ended intervals, say $(-\infty, 4.75)$ and $(9.25, \infty)$. In between these, use intervals of width .9: 4.8-5.6, 5.7-6.5, etc.]

Depth	Stm	Leaves
1	3	1
1	3	
1	4	
2	4	5
5	5	443
10	5	66669
16	6	011114
33	6	56666666677889999
13	7	1111112233444
26	7	566688888899
14	8	033345
8	8	5668
4	9	22
2	9	9
1	10	
1	10	9

$n = 72$

Sections 13.3-13.4

13-C Problem 7-D gave these percentages of nitrogen oxides removed in a coal-burning facility:

91 95 90 83 91 65 55 42 55 81 89 38 20 45 58 85 78 70.

Given: $\bar{X} = 68.39$, $S = 21.54$.
(a) Test the hypothesis that the distribution is $N(75, 15^2)$, using the K-S statistic.
(b) Test the hypothesis that the population is normal, using the Lilliefors test.
(c) Construct a rankit plot and calculate the Shapiro-Wilk statistic.

13-D For the data in Problem 13-B, we made a rankit plot using computer software and found $W = .97$. Find the P-value, for testing normality of the population.

Section 13.5

13-E A study of the dramatic increase in the number of working women gave these data on husbands' attitudes:[7]

	Husbands with employed wives	Husbands with nonemployed wives
n	341	141
Like wives working	74%	37%
Do not care	11%	26%
Dislike wives working	15%	37%

Is the pattern of attitudes different between the two types of husbands?

13-F A study considered 180 Harvard medical students who contacted a mental health service over a period of 5 years.[8] Of the 85 women in the study, 69 selected a psychiatrist of the same sex; of the 95 men, 62 selected a psychiatrist of the same sex. How strong is the evidence that men and women differ in the matter of selecting a psychiatrist of the same sex,
(a) using a Z-test?
(b) using a chi-square test?

---

[7]D. V. Hiller and W. Philliber, "The division of labor in contemporary marriage: Expectations, perceptions, and performance," *Social Problems* (1986), 191–201.

[8]K. B. Kris and J. Silberger, "Medical students' gender preference when selecting a psychiatrist," *J. Amer. College Health* **35** (1986), 5–10.

Section 13.6–13.7

13-G Carry out a likelihood ratio test of homogeneity with the data in Problem 13-F above.

13-H Do traditional female roles tend to conflict with out-of-town travel? A study reported these data on willingness to travel for a job:[9]

	Unwilling	Willing, a week or two	Willing, a month or two
Males	74	78	339
Females	81	127	199

Use a likelihood ratio test.

[9]W. T. Markham, C. M. Bonjean, and J. Corder, "Gender, out-of-town travel, and occupational advancement," *Sociology and Social Research* (1986), 156–160.

Chapter 13: Solutions

Section 13.1

13-A Solution:

The sample size is 518. The observed and expected frequencies and the differences between observed and expected frequencies are as follows:

Color	Observed	Expected	Difference
Brown	234	207.2	26.8
Yellow	101	103.6	−2.6
Orange	91	103.6	−12.6
Tan	50	51.8	−1.8
Green	42	51.8	−9.8

Substituting in (1) of §13.2 we get

$$\chi^2 = \frac{26.8^2}{207.2} + \frac{2.6^2}{103.6} + \frac{12.6^2}{103.6} + \frac{1.8^2}{51.8} + \frac{9.8^2}{51.8} \doteq 6.98.$$

There are 5 categories, so the distribution of χ^2 under H_0 is chi$^2(5-1)$. Table Vb shows that the P-value is greater than .126.

13-B Solution:

Using the specified class-marks, the frequency table is shown below. The sample mean and s.d. are $\bar{X} = 7.125$, $S = 1.2214$. With the given scheme, there will be too many class intervals with small expected frequencies where the data thin out at each end, so we'll use intervals $(-\infty, 4.75)$ at the left and $(9.25, \infty)$ at the right.

We need z-scores for the dividing points: 4.75, 5.65, 6.55, etc., in order to calculate probabilities of the categories (class intervals). Since we don't have population parameters, we approximate the z-scores using the sample moments. Thus, the probability of the first interval is

$$P(X < 4.75) = \Phi\left(\frac{4.75 - 7.125}{1.2214}\right) = .0259.$$

For the second,

$$P(4.75 < X < 5.65) = \Phi\left(\frac{5.65 - 7.125}{1.2214}\right) - .0259 = .0873.$$

And so on. The result is the following table in which we show observed frequencies, interval probabilities, and expected frequencies:

Class interval	Frequency	p_i	np_i
-4.7	2	.026	1.9
4.8-5.6	7	.087	6.3
5.7-6.5	8	.205	14.8
6.6-7.4	29	.288	20.7
7.5-8.3	16	.237	17.1
8.4-9.2	8	.117	8.4
9.3-	2	.041	2.9

Finally, substitute observed and expected frequencies in the formula for χ^2 (using a hand calculator) to obtain $\chi^2 = 6.9$. There are 7 categories and 2 estimated parameters, so d.f. $= 7 - 1 - 2 = 4$. The P-value is on the order of .15 (interpolating in Table Va). This is not significant, but even with 72 observations, the chi-square test is not all that powerful. (Just looking at the stem-leaf diagram suggests that maybe the tails are too fat for normality; but the observations in the tails were rounded into semi-infinite intervals, and the tail frequencies would be the same if the most extreme observations had been 4.4 and 9.9 instead of 3.1 and 10.9.)

13-C Solution:
(a) Figure 9 shows the sample c.d.f. together with the normal c.d.f. of the H_0 distribution. The maximum deviation occurs at $x = 58$:

$$D_n = \frac{7}{18} - \Phi\left(\frac{58 - 75}{15}\right) \doteq .26.$$

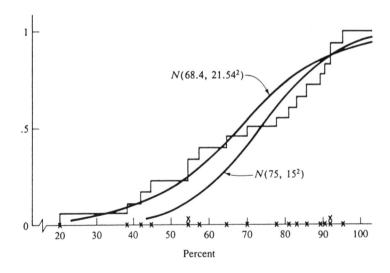

Figure 9

(b) The c.d.f. of $N(68.39, 21.54^2)$ is also shown in Figure 9. The maximum deviation is 17, at $x = 78$. The 20% critical value from Table XI is .167, so $P > .20$.

(c) The rankit plot, shown in Figure 10, is a scatter plot of the pairs $(X_{(i)}, E[Z_{(i)}])$, where the normal scores $E[Z_{(i)}]$ are found in Table XII. The Wilk-Shapiro statistic, given by our computer software, is .927. The P-value appears to be just over .10, from Table XIV. To calculate this statistic, you first find the rankits in Table XII, $n = 18$: -1.83, -1.35, -1.066, $-.848$, etc., and pair these with the *ordered* observations: 20, 38, 42, 45, etc. Then $W = r^2$, where r is the correlation coefficient of the rankits and the ordered observations.

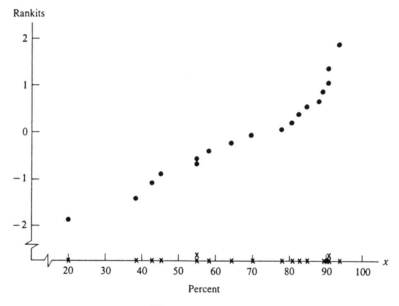

Figure 10

13-D Solution:

Table XIV gives the 5th percentile of W for $n = 75$ as .969, and the 5th percentile when $n = 72$ will be slightly less. So with $W = .97$, the P-value is just a little larger than .05, perhaps .06—marginal significance.

13-E Solution:

To use χ^2 we need frequencies. Multiply each given percentage by the appropriate sample size $341 \times .74 = 252$, $141 \times .37$, etc., to obtain the frequencies in the first table below:

252	52	304
38	37	75
51	52	103
341	141	482

215.1	86.9	304
53.1	21.9	75
72.8	30.2	103
341	141	482

The expected frequencies in the second table are found using (4) of §13.5: $341 \times 304/482 \doteq 215.1$, $341 \times 75/482 \doteq 53.1$, $141 \times 304/482 \doteq 88.9$, etc. Then, using the chi-square formula (5), find

$$\chi^2 = \frac{36.9^2}{215.1} + \frac{36.9^2}{88.9} + \frac{15.1^2}{53.1} + \frac{15.1^2}{21.9} + \frac{21.8^2}{72.8} + \frac{21.8^2}{30.2} \doteq 58.6.$$

Use Table Vb with d.f. $= (3-1)(2-1) = 2$: $P \ll .002$. The evidence is very strong against the hypothesis of homogeneity.

13-F Solution:

(a) The Z-score for comparing sample proportions is

$$Z = \frac{\frac{69}{85} - \frac{62}{95}}{\sqrt{\frac{131}{180} \cdot \frac{49}{180}\left(\frac{1}{85} + \frac{1}{95}\right)}} = \frac{.1591}{.06645} = 2.395.$$

The tail probability (from Table II) is about .0083—rather strong evidence against H_0.

(b) For the chi-square test we need estimates of the expected frequencies, from (4) of §13.5: $131 \times 85/180 \doteq 61.9$. Subtracting from row and column totals yields the others: 23.1, 69.1, 25.9. In the 2×2 case, all differences between expected and observed frequencies are the same, in this case, 7.1. Then

$$\chi^2 = 7.1^2\left(\frac{1}{61.9} + \frac{1}{23.1} + \frac{1}{69.1} + \frac{1}{25.9}\right) \doteq 5.7.$$

This is the square of Z in (a)—exactly so, if we had not rounded off. But the P-value from the chi-square table is .017—twice that in (a), because the chi-square test is automatically two-sided.

13-G Solution:

All we need do is substitute the joint frequencies, the marginal frequencies, and the sample sizes into (6) of §13.6:

$$\log \Lambda = 131 \log 131 + 49 \log 49 + 85 \log 85 + 95 \log 95$$

$$- 69 \log 69 - 62 \log 62 - 16 \log 16 - 33 \log 33 - 180 \log 180 = -2.92,$$

so $-2 \log \Lambda = 5.84$. (Observe that this is close to χ^2 in 13-F.)

13-H   Solution:

Again, we simply substitute the given data and $n = 898$ in (6) of §13.6:

$$-\log \Lambda = 898 \log 898 + 74 \log 74 + 78 \log 78 + 339 \log 339$$

$$+ 81 \log 81 + 127 \log 127 + 199 \log 199 - 155 \log 155$$

$$- 205 \log 205 - 538 \log 538 - 491 \log 491 - 407 \log 407 = 20.56.$$

Then $-2 \log \Lambda = 41.4$. The largest entry in Table Vb for χ^2 is 13, for which $P = .002$, so P is much smaller than .002.

CHAPTER 14: Additional Problems

Sections 14.1-14.2

14-A Show that SSTot and SSTr in 1-way ANOVA can be calculated as follows:

(a) $\text{SSTot} = \sum X_{ij}^2 - \frac{1}{n}\left(\sum X_{ij}\right)^2.$

(b) $\text{SSTr} = \sum n_i \bar{X}_i^2 - \frac{1}{n}\left(\sum X_{ij}\right)^2.$

14-B A tensile strength test measures the quality of a spot-weld. Five welds of an aluminum-clad material done with each of three machines were tested, with these results:

Machine A: 3.2, 4.1, 3.5, 3.0, 3.1 (mean = 3.38)
Machine B: 4.9, 4.5, 4.5, 4.0, 4.2 (mean = 4.42)
Machine C: 3.0, 2.9, 3.7, 3.5, 4.2 (mean = 3.46)

(a) Test the hypothesis of no "machine effect," using the formulas in (a).
(b) Recalculate the error sum of squares, given these sample standard deviations: .443, .342, .532 for A, B, and C, respectively.
(c) Construct a 90 percent confidence interval for $\mu_B - \mu_A$, the difference in mean strengths for machines A and B.

Section 14.3

14-C Given the data in Problem 14-B, find statistically significant differences when $\alpha = .05$ using
(a) the Bonferroni method.
(b) the Scheffé method.

Section 14.4

14-D Three machines are to be compared with respect to output over a given time period. Since the output may depend on the operator as well as the machine, an experiment measured the output for each of four operators on each of the three machines:

	Operator			
	34	35	38	33
Machine:	30	43	40	31
	32	36	39	29

Test the hypothesis of no (average) difference among machine outputs.

14-E The data on muscle stimulation of lambs' trachea from Example 12.6a are repeated here (multiplied by 1000 to simplify the arithmetic). Treat these paired data as from a randomized block design, where the size of the block is $c = 2$, and test for a treatment effect:

Lamb #	Before	After
1	29	20
2	43	11
3	22	8
4	12	5
5	20	9
6	34	9

14-F In the notation and context of §14.4, show the following:

$$E\left(\sum n_i(\bar{X}_i. - \bar{X})^2\right) = \frac{n(r-1)\sigma^2}{rc} + \sum n_i\tau_i^2.$$

(This shows that SSTr tends to be larger when the treatment effects τ_i are not all 0 than when they are.)

Chapter 14: Solutions

Section 14.1

14-A Solution:

(a) This formula is simply the parallel axis theorem for the sample consisting of all the data together [(5) in §7.5].

(b) This is derived in much the same way as we derived the parallel axis theorem. First, we expand the square in the defining formula:

$$\text{SSTr} = \sum n_i (\bar{X}_i - \bar{X})^2 = \sum n_i (\bar{X}_i^2 - 2\bar{X}\bar{X}_i + \bar{X}^2)^2.$$

Sum term by term:

$$\text{SSTr} = \sum n_i \bar{X}_i^2 - 2\bar{X} \sum n_i \bar{X}_i + \bar{X}^2 \sum n_i,$$

and recall that $n\bar{X} = \sum n_i \bar{X}_i$, and the sum of the n_i is n:

$$\text{SSTr} = \sum n_i \bar{X}_i^2 - n\bar{X}^2$$

Now substitute $\bar{X} = \frac{1}{n} \sum \sum X_{ij}$, cancel an n, and we're home free.

14-B Solution:

(a) The sum of all the observations is 56.3. The row averages are 3.38, 4.42, 3.46, with sum of squares equal to 42.756. Then,

$$\text{SSTr} = 5 \times 42.9324 - \tfrac{1}{15} \times 56.3^2 = 3.3493.$$

and

$$\text{SSTot} = 217.05 - \tfrac{1}{15} \times 56.3^2 = 5.7373.$$

The error sum of squares is the difference, $5.7373 - 3.3493 = 2.388$, and

$$F = \frac{\text{MSTr}}{\text{MSE}} = \frac{3.349/2}{2.388/12} = 8.42.$$

With 2 and 12 d.f., we find $P \doteq .005$ (in Table VIIIa).

(b) $\text{SSE} = \sum (n_i - 1)S_i^2 = 4(.443^2 + .342^2 + .532^2) = 2.385$, differing from the above only because of round-off.

(c) The confidence interval will be of the form $\bar{X}_B - \bar{X}_A \pm k \cdot \text{s.e.}$, where k is a t-percentile. The mean difference is 1.04, and we'll use the usual formula for standard error, except that we can use the pooled variance estimate based on all 3 samples: $S_p^2 = \text{MSE} = 2.388/12 = .199$

$$\text{s.e.} = \sqrt{\text{MSE}\left(\tfrac{1}{5} + \tfrac{1}{5}\right)} = .282.$$

For 90% confidence, we use the 95th percentile of $t(8)$: $k = 1.86$. The desired interval is $1.04 \pm 1.86 \times .282 = 1.04 \pm .525$.

14-C **Solution:**

(a) In (c) of the preceding problem we found one confidence interval. Here we want three confidence intervals with a simultaneous error rate of .05. As in Example 14.3b, we make three comparisons: A with B, A with C, and B with C. The sample sizes are the same, so the standard errors will be the same for each comparison, namely, the s.e. found in Problem 14-B: .0796. The multiplier is found using $\alpha/m = .05/3 = .0083$, but here the number of degrees of freedom in the variance estimate is $15 - 3 = 12$. The value of t for a two-tail area of $2 \times .0083$, from Table IIIb, is about 2.8. So, for confidence limits, we go either way, from the sample mean, an amount $2.8 \times .282 \doteq .79$. The confidence limits are

$$\text{For } \mu_B - \mu_A: \ 1.04 \pm .79$$
$$\text{For } \mu_B - \mu_C: \ .96 \pm .79$$
$$\text{For } \mu_C - \mu_A: \ .08 \pm .79.$$

The first two intervals do not contain 0, so the means of populations B and A, and of populations B and C are judged to be different; the means of populations C and A are not.

(b) Following the pattern of Example 14.3c, we first find the 95th percentile of $F(2, 12)$: 3.89. The quantity "b" as defined by (6) in §14.3 is

$$b = (3 - 1) \times 3.89 \times (1/5 + 1/5) = 3.11.$$

The quantity $S_p\sqrt{b}$ in (5) of that section is then $\sqrt{.199 \times 3.11} \doteq .79$, so the intervals will differ only in round-off from those in (a).

14-D **Solution:**

You can use the formulas in Problem 14-22 to find the various sums of squares, but they're not hard to find from their basic definitions. The row means are 35, 36, 34, and the grand mean is 35. So

$$\text{SSTr} = 4[(35 - 35)^2 + (36 - 35)^2 + (34 - 35)^2] = 8.$$

The column means are 32, 38, 39, 31, so

$$\text{SSB} = 3[(32 - 35)^2 + (38 - 35)^2 + (39 - 35)^2 + (31 - 35)^2] = 150.$$

The total sum of squares is the sum of the deviations of the 12 entries about the grand mean: $(34 - 35)^2 + (35 - 35)^2 + \cdots = 206$. From these we can get the error sum of squares by subtraction: $206 - 150 - 8 = 48$. The ANOVA table is as follows:

	df	SS	MS	F
Machine	2	8	4	.50
Operator	3	150	50	6.25
Error	6	48	8	

The F-ratio for machine effect is not significant; the 6.25, with (3, 6) df, gives a P-value of .028 (Table VIIIa).

14-E   Solution:

We use the formulas given in Problem 14-22 to find the sums of squares using a hand calculator:

$$C = \tfrac{1}{rc}\left(\sum\sum X_{ij}\right)^2 = \tfrac{1}{12}\times 222^2 = 4107.$$

The sums in the before and after columns are 160 and 62, so

$$\text{SSTr} = \tfrac{1}{6}(160^2 + 62^2) - C = 4907.33 - 4107 = 800.33.$$

The block sums 49, 54, 30, 17, 29, 43. Squaring, summing, and dividing by 2 yields 4598, so

$$\text{SSB} = 4598 - 4107 = 491.$$

And SSTotal is just the numerator of the variance of all the observations, which we can calculate as

$$\sum\sum X_{ij}^2 - rc\bar{X}^2 = \sum\sum X_{ij}^2 - C = 5646 - 4107 = 1539.$$

The error sum of squares is found by subtraction: $1539 - 491 - 800 = 248$. So the F-ratio for testing the hypothesis of no treatment effect is

$$F_{\text{Tr}} = \frac{800.33/1}{248/5} = 16.136.$$

In the example referred to, we found $T = 4.06$, which is almost the square root of F (differing because of round-off).

14-F   Solution:

This is not trivial—don't be surprised if you got stuck trying to do it! But try to follow these steps:

We expand the following square:

$$(\bar{X}_i. - \bar{X} - \tau_i)^2 = (\bar{X}_i. - \bar{X})^2 + \tau_i^2 - 2\tau_i(\bar{X}_i. - \bar{X}).$$

Multiplying by n_i and summing on i, we get

$$\sum n_i(\bar{X}_i. - \bar{X} - \tau_i)^2 = \sum n_i(\bar{X}_i. - \bar{X})^2 + \sum n_i\tau_i^2 - 2\sum n_i\tau_i(\bar{X}_i. - \bar{X}).$$

Upon taking expected values, we find that

$$E[2\sum n_i\tau_i(\bar{X}_i. - \bar{X})] = 2\sum n_i\tau_i E[(\bar{X}_i. - \bar{X})] = 2\sum n_i\tau_i^2.$$

Thus,

$$\sum n_i E[(\bar{X}_i. - \bar{X} - \tau_i)^2] = \sum n_i E[(\bar{X}_i. - \bar{X})^2] - \sum n_i\tau_i^2,$$

or

$$\text{SSTr} = \sum n_i E[(\bar{X}_i. - \bar{X} - \tau_i)^2] + \sum n_i\tau_i^2. \tag{1}$$

It remains only to evaluate the first term on the right. Since

$E(\bar{X}_{i\,.} - \bar{X}) = \tau_i$, the expected square of $\bar{X}_{i\,.} - \bar{X} - \tau_i$ is its variance:

$$\operatorname{var}(\bar{X}_{i\,.} - \bar{X}) = \operatorname{var}\bar{X}_{i\,.} + \operatorname{var}\bar{X} - 2\operatorname{cov}(\bar{X}_{i\,.}, \bar{X}). \tag{2}$$

The first term is σ^2/c, and the second is $\sigma^2/(rc)$. And expressing $\bar{X}$ as the average of the row means, we have

$$\operatorname{cov}(\bar{X}_{i\,.}, \bar{X}) = \operatorname{cov}(\bar{X}_{i\,.}, \tfrac{1}{r}\sum_1^r \bar{X}_{k\,.}),$$

where we have used the running index k instead of i to avoid confusion with the earlier subscript i. Since all the data are independent, the data in row i are independent of the data in other rows, and in particular, independent of $\bar{X}_{k\,.}$, except when $k = i$. Thus,

$$\operatorname{cov}(\bar{X}_{i\,.}, \bar{X}) = \operatorname{cov}(\bar{X}_{i\,.}, \tfrac{1}{r}\bar{X}_{i\,.}) = \tfrac{1}{r}\operatorname{var}(\bar{X}_{i\,.}) = \tfrac{\sigma^2}{rc}.$$

Going back to (2), we see that

$$\operatorname{var}(\bar{X}_{i\,.} - \bar{X}) = \sigma^2\Big(\tfrac{1}{c} + \tfrac{1}{rc} - 2\cdot\tfrac{1}{rc}\Big) = \frac{(r-1)\sigma^2}{rc}.$$

And substituting this in (1) for $E[(\bar{X}_{i\,.} - \bar{X} - \tau_i)^2]$ yields the formula we were asked to obtain.

CHAPTER 15: Additional Problems

Sections 15.1-15.2

15-A For the data in Example 15.1a, find
(a) the correlation coefficient.
(b) the equation of the least-squares regression line.

15-B Given a least-squares line $\hat{\alpha} + \hat{\beta} x$ (in the notation and setting of these sections), let $\hat{Y}_i = \hat{\alpha} + \hat{\beta} x_i$, the "fitted value" at x_i, Show that in the n pairs $(Y_i, \hat{Y}_i)$, the correlation coefficient is $|r|$, where r is the correlation coefficient of the data pairs (x_i, Y_i).

15-C Use the method of least squares to fit a quadratic function through the origin to these three data points: $(1, 0)$, $(2, 2)$, $(3, 2)$.

Section 15.3-15.4

15-D In §15.3 [page 603] we say that the cross-product terms in (8) vanish. Show that they do.

15-E Show the following identity, asserted to hold in §15.4 (page 607):

$$n\hat{\sigma}_0^2 = \text{SSRes} + (\hat{\beta} - \beta_0)^2 \text{SS}_{xx},$$

where $\hat{\sigma}_0^2$ is the m.l.e. of the error variance when $\beta = \beta_0$ in the simple linear regression model:

$$\hat{\sigma}_0^2 = \tfrac{1}{n}\sum (Y_i - \hat{\alpha}_0 - \beta_0 x_i)^2, \quad \hat{\alpha}_0 = \bar{Y} - \beta_0 \bar{x}.$$

15-F Problem 15-14 deals with the estimation of the mean response at a new value x_0, based on the data in n cases, using the estimated mean, $\hat{\alpha} + \hat{\beta} x_0$. The mean and variance of this estimator $\hat{Y}$ are given in Problem 15-14. Obtain confidence for the true mean response at x_0, assuming the normal model for simple linear regression (as in §15.2-15.4).

Sections 15.5-15.7

15-G Suppose we find that the correlation between first and second exam scores in a particular course is $\rho = .90$, and that the mean scores are 65 and the s.d.'s are 12. Predict the second exam score, and give the corresponding

r.m.s. prediction error, of a student whose first exam score is
(a) 45.   (b) 95.   (c) unknown.

15-H   Given the data in Example 15.1a [which are repeated in the solution to
Problem 15-A]. Give a predicted value of Y when $x = 3.5$, together with
an estimate of the r.m.s. prediction error and a 95% prediction interval.

15-I   Observe and comment on the regression effect, in Problem 15-G.

Section 15.8

15-J   Eye weight in grams and corneal thickness in micrometers were recorded
for nine randomly selected calves:[10]

Calf	1	2	3	4	5	6	7	8	9
Weight (x)	.2	1.4	2.2	2.7	4.9	5.3	8.0	8.8	9.6
Thickness (y)	416	673	733	801	967	1036	883	736	567

(a) Make a scatter plot. Would a linear regression function be a good fit?
(b) Write in mathematical form the hypothesis that the regression
function for y on x is quadratic against the hypothesis that it is linear.
(c) Using computer software, obtain the least-squares quadratic regression.
(d) Test the hypothesis in (b).
(e) Obtain a residual plot.

---

[10]Data are from "Collagens of the developing bovine cornea," *Exper. Eye Res.*
(1984), 639-652.

Chapter 15: Solutions

Section 15.1

15-A   Solution:

The data are shown in the accompanying table, along with columns for finding the sums we need for r and the regression line:

x	y	x^2	xy	y^2
2	89	4	178	7921
2.5	97	6.25	242.5	9409
2.5	91	6.25	227.5	8281
2.75	98	7.5625	269.5	9604
3	100	9	300	10000
3	104	9	312	10816
3	97	9	291	9409
18.75	676	51.0625	1820.5	65440

(a) To find r, we use (3) of §7.6:

$$r = \frac{1820.5 - \dfrac{18.75 \times 676}{7}}{\sqrt{51.0625 - \dfrac{(18.75)^2}{7}}\sqrt{65440 - \dfrac{676^2}{7}}} = \frac{9.7857}{\sqrt{.8393} \times 157.714} \doteq .85.$$

(b) The slope is SS_{xy} (the numerator of r) divided by SS_{xx} (which is under the first square root sign in the denominator of r above):

$$\hat{\beta} = \frac{9.7857}{.8393} = 11.66.$$

The intercept is $\hat{\alpha} = \bar{Y} - \hat{\beta}\bar{x} = 96.57 - 11.66 \times 2.679 = 65.34$. So the equation of the line is $y = 65.34 + 11.66\,x$.

15-B   Solution:

The average of the fitted values is $\bar{Y}$:

$$\tfrac{1}{n}\sum \hat{Y}_i = \tfrac{1}{n}\sum(\hat{\alpha} + \hat{\beta}\,x_i) = \hat{\alpha} + \hat{\beta}\bar{x} = \bar{Y}.$$

And

$$\hat{Y}_i - \bar{Y} = \hat{\alpha} + \hat{\beta}\,x_i - (\hat{\alpha} + \hat{\beta}\bar{x}) = \hat{\beta}(x_i - \bar{x}).$$

Thus,

$$SS_{\hat{y}\hat{y}} = \sum(\hat{Y}_i - \bar{Y})^2 = \hat{\beta}^2\,SS_{xx},$$

and

$$SS_{y\hat{y}} = \sum Y_i(\hat{Y}_i - \bar{Y}) = \hat{\beta}\sum Y_i(x_i - \bar{x}) = \hat{\beta}\,SS_{xy}.$$

So, the correlation of Y_i with $\hat{Y}_i$ is

$$r_{y\widehat{y}} = \frac{SS_{y\widehat{y}}}{\sqrt{SS_{yy}SS_{\widehat{y}\widehat{y}}}} = \frac{\widehat{\beta}\, SS_{xy}}{\sqrt{SS_{yy}\cdot\widehat{\beta}^2\,SS_{xx}}} = \frac{\widehat{\beta}\, SS_{xy}}{|\widehat{\beta}|\sqrt{SS_{xx}SS_{yy}}} = |r|.$$

15-C Solution:

The residual about $\alpha x + \beta x^2$ (no constant term, so that the graph goes through the origin) is $y - (\alpha x + \beta x^2)$. Substituting the coordinates of the data points, in turn, and squaring we get

$$\sum(y - \alpha x_i - \beta x_i^2)^2 = (-\alpha - \beta)^2 + (2 - 2\alpha - 4\beta)^2 + (2 - 3\alpha - 9\beta)^2.$$

To maximize, we differentiate with respect to each variable in turn and set the derivatives equal to 0:

$$2(\alpha + \beta) + 2(2 - 2\alpha - 4\beta)(-2) + 2(2 - 3\alpha - 9\beta)(-3) = 0$$

$$2(\alpha - \beta) + 2(2 - 2\alpha - 4\beta)(-4) + 2(2 - 3\alpha - 9\beta)(-9) = 0.$$

Collecting terms and simplifying, we get

$$\begin{cases} 14\alpha + 36\beta = 10 \\ 36\alpha + 98\beta = 26. \end{cases}$$

Solve simultaneously to get the desired coefficients: $\alpha = 11/19$, $\beta = 1/19$. The equation is $Y = \frac{11}{19}x + \frac{1}{19}x^2$.

15-D Solution:

The three cross products in the square of the trinomial are

$$(Y_i - \widehat{\alpha} - \widehat{\beta}x_i)[\bar{Y} - (\alpha + \beta\,\bar{x})], \tag{1}$$

$$(\widehat{\beta} - \beta)(x_i - \bar{x})[\bar{Y} - (\alpha + \beta\,\bar{x})], \tag{2}$$

$$(Y_i - \widehat{\alpha} - \widehat{\beta}x_i)(\widehat{\beta} - \beta)(x_i - \bar{x}). \tag{3}$$

In summing (1) and (2) on i, the bracketed factor is constant, and $\widehat{\beta} - \beta$ is constant. So the result of summing (2) on i is a constant multiplied by $\sum(x_i - \bar{x})$, which is 0. Summing (1) on i also yields a 0, according to one of the two equations to be solved in Problem 15-5 (to obtain the least squares coefficients:

$$\sum(Y_i - \widehat{\alpha} - \widehat{\beta}x_i) = \sum(Y_i - \bar{Y} + \widehat{\beta}\bar{x} - \widehat{\beta}x_i) = \sum(Y_i - \bar{Y}) + \widehat{\beta}\sum(x - \bar{x}).$$

And the sums of deviations (Y_i's about their mean, and x_i's about theirs) are both 0.

That the sum of expressions (3) also vanishes is essentially the second of those two equations which emerge in doing Problem 15-5:

$$\sum(Y_i - \widehat{\alpha} - \widehat{\beta}x_i)(x_i - \bar{x}) = \sum x_i(Y_i - \widehat{\alpha} - \widehat{\beta}x_i) = 0.$$

In going from the left side to the right, we drop the $\bar{x}$; in so doing, we really drop nothing, since $\bar{x}$ is multiplied by $(Y_i - \widehat{\alpha} - \widehat{\beta}x_i)$, and these sum to 0, as in (1). To show this, we multiply out the product:

$$\sum (x_i - \bar{x})Y_i - \hat{\alpha}\sum (x_i - \bar{x}) - \hat{\beta}\sum x_i(x_i - \bar{x}) = SS_{xy} - 0 - \hat{\beta} SS_{xx},$$

and find that this vanishes, because we defined $\hat{\beta}$ as SS_{xy}/SS_{xx}.

15-E   Solution:
In the terms summed to get $\hat{\sigma}_0$, we substitute $\bar{Y} - \beta_0\bar{x}$ for $\hat{\alpha}_0$, add and subtract $\hat{\beta}(x_i - \bar{x})$, and rearrange terms:

$$Y_i - \hat{\alpha}_0 - \beta_0 x_i = Y_i - \bar{Y} + \beta_0\bar{x} - \beta_0 x_i$$
$$= [Y_i - \bar{Y} - \hat{\beta}(x_i - \bar{x})] + (\hat{\beta} - \beta_0)(x_i - \bar{x})$$
$$= (Y_i - \hat{\alpha} - \hat{\beta}x_i) + (\hat{\beta} - \beta_0)(x_i - \bar{x}).$$

Next, square this as a binomial, and sum:

$$n\hat{\sigma}_0^2 = \sum (Y_i - \hat{\alpha} - \hat{\beta}x_i)^2 + (\hat{\beta} - \beta_0)^2 \sum (x_i - \bar{x})^2.$$

The cross-product is 0 because of what we showed in Problem 15-D:

$$\sum (Y_i - \hat{\alpha} - \hat{\beta}x_i)(\hat{\beta} - \beta_0)(x_i - \bar{x}) = (\hat{\beta} - \beta_0) \sum x_i(Y_i - \hat{\alpha} - \hat{\beta}x_i) = 0.$$

Now, since $\hat{\alpha} = \bar{Y} - \hat{\beta}\bar{x}$, we can write the first term on the right of $n\hat{\sigma}_0^2$:

$$\sum [Y_i - \bar{Y} - \hat{\beta}(x_i - \bar{x})]^2 = \sum (Y_i - \hat{\alpha} - \hat{\beta}\bar{x})^2 = SSRes.$$

But $\sum (x_i - \bar{x})^2 = SS_{xx}$, so this completes the proof.

15-F   Solution:
Because $\hat{Y}$ is normal, the square of the standardized $\hat{Y}$ has a $\text{chi}^2(1)$ distribution:

$$\left\{ \frac{\hat{Y} - E\hat{Y}}{\sigma_{\hat{Y}}} \right\}^2 \sim \text{chi}^2(1),$$

where

$$\sigma_{\hat{Y}} = \sigma \sqrt{\frac{1}{n} + \frac{(x_0 - \bar{x})^2}{SS_{xx}}}.$$

We get the standard error of $\hat{Y}$ by replacing σ by S_e [(11) in §15.3]

$$s.e.(\hat{Y}) = S_e \sqrt{\frac{1}{n} + \frac{(x_0 - \bar{x})^2}{SS_{xx}}}.$$

From §15.3 (page 604) we know that

$$\frac{(n-2)S_e^2}{\sigma^2} \sim \text{chi}^2(n-2)$$

and is independent of $\hat{Y}$. Now define

$$T = \frac{\widehat{Y} - (\alpha + \beta x_0)}{\text{s.e.}(\widehat{Y})} = \frac{\widehat{Y} - (\alpha + \beta x_0)}{S_e \sqrt{\frac{1}{n} + \frac{(x_0 - \bar{x})^2}{SS_{xx}}}}.$$

Dividing the numerator and denominator in the last expression by σ yields a numerator which is standard normal, and a denominator which is the square root of a chi-square variable divided by its d.f. Its distribution is thus $t(n-2)$, so it is pivotal. The resulting confidence limits for $\alpha + \beta x_0$ are $\widehat{Y} \pm k[\text{s.e.}(\widehat{Y})]$, where k is an appropriate percentile of $t(n-2)$, found in Table IIIa as determined by the confidence level.

15-G   Solution:

We use the linear predictor (8) of §15.5 in its equivalent form given by (3) of §15.5:

$$y = \mu_Y + \rho \frac{\sigma_Y}{\sigma_X}(x - \mu_X).$$

(a) Set $x = 45$ and use the given means and variances:

$$y = 65 + .9(45 - 65) = 47.$$

The r.m.s.p.e. is $\sigma_Y \sqrt{1 - \rho^2} = 12\sqrt{1 - .81} = 5.23$.

(b) Set $x = 95$: $y = 65 + .9(95 - 65) = 92$, with the same r.m.s.p.e., 5.23.

(c) With the first exam score unknown, we predict Y to have its mean value, 65, with r.m.s. prediction error equal to its standard deviation, 12.

15-H   Solution:

We found the least squares line in Problem 15-A: $y = 65.34 + 11.66\,x$, and we substitute $x = 3.5$ to find the predicted value: $\widehat{Y} = 106.15$. To get at the prediction error, we need to estimate σ_e, which we do using

$$S_e = \sqrt{\text{MSE}} = \sqrt{\frac{\text{SSRes}}{n-2}} = \sqrt{\frac{SS_{yy}(1 - r^2)}{n-2}} = \sqrt{\frac{157.714(1 - .72343)}{7 - 2}} = 2.954$$

(taking r and SS_{yy} from Problem 15-A). We then use this in (9) of §15.5, to estimate the r.m.s. prediction (with $\bar{x}$ and SS_{xx} from Problem 15-A):

$$2.954\sqrt{1 + \frac{1}{7} + \frac{(3.5 - 2.697)^2}{.8393}} \doteq 4.12.$$

For a 95% prediction interval, we multiply this s.e. of prediction by the 97.5 percentile of $t(5)$: 2.57, to obtain the desired limits 106.15 ± 10.59.

15-I   Solution:

The predicted 2nd exam score for a student with a low 1st exam score (45) is higher—closer to the mean; the predicted score of a student with a high score (95) is not so high—closer to average. This is the "regression effect" in action.

15-J   Solution:

(a) The scatter plot (shown below) looks more quadratic than linear.

(b) The quadratic model says that the response at x has mean value $\alpha + \beta_1 x + \beta_2 x^2$. The null hypothesis is $\beta_2 = 0$, against the alternative $\beta_2 \neq 0$—but it is more precise to say that H_0 is the hypothesis $EY_x = \alpha + \beta_1 x$, against the hypothesis that $EY_x = \alpha + \beta_1 x + \beta_2 x^2$, where $\beta \neq 0$.

(c) Our software produced the coefficient and ANOVA tables shown below.

(d) The t-statistic for the hypothesis in (b) is shown opposite "XSQ" in the coefficient table: $T = -3.64$ (d.f. $= 6$), with $P = .0108$. This is strong evidence against $\beta_2 = 0$ (provided x_1 is left in the model), as you would expect looking at the scatter plot.

(e) The standardized residuals are plotted below.

PREDICTOR VARIABLES	COEFFICIENT	STD ERROR	STUDENT'S T	P	VIF
CONSTANT	349.601	71.8229	4.87	0.0028	
X	245.127	38.4779	6.37	0.0007	19.7
XSQ	-23.6277	3.91144	-6.04	0.0009	19.7

R-SQUARED	0.8727	RESID. MEAN SQUARE (MSE)	6370.15
ADJUSTED R-SQUARED	0.8302	STANDARD DEVIATION	79.8132

SOURCE	DF	SS	MS	F	P
REGRESSION	2	2.620E+05	1.310E+05	20.56	0.0021
RESIDUAL	6	38220.9	6370.15		
TOTAL	8	3.002E+05			

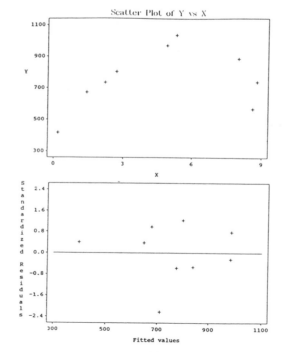

Part II

Solutions to Review Problems

CHAPTER 1

1-R1 (a) $(E \cup F)^c EF = (E^c F^c) EF = E^c E F^c F = \emptyset$.

   (b) Applying the distributive law, we have
$$(E \cup F)(E \cup F^c) = EE \cup EF^c \cup FE \cup FF^c$$
$$= E \cup E(F \cup F^c) \cup \emptyset = E \cup E \cup \emptyset = E$$

1-R2 (a) The left-hand side is $A(B - C) = ABC^c$. We work with the right side:
$$AB - AC = AB(AC)^c = AB(A^c \cup C^c) = ABA^c \cup ABC^c = ABC^c = \text{l.h.s.}$$
   (b) $A = B \cup AB^c$, so $P(A) = P(B) + P(AB^c) = P(B) + P(A - B)$.

1-R3 This is just the number of arrangements of 6 things in a sequence: 6!

1-R4 Choose 4 from 12 for A, then 4 from the remaining 8 for B; the rest go to B: $\binom{12}{4}\binom{8}{4} = 34650$. This is also $\binom{12}{4, 4, 4} = \frac{12!}{4!4!4!}$. [See (6), §1.4.]

1-R5 It helps to draw a picture. Represent A as the circle, B as the triangle:

The complement of $A \cup B^c$ is shaded. It is $(A \cup B^c)^c = A^c B$. The part of B not in AB has to have probability $.3 - .21 = .09$, so $P(A \cup B^c) = 1 - .09 = .91$. Or, alternatively, we can use the formula for the probability of a union, with $P(AB^c) = P(A) - P(AB) = .29$:

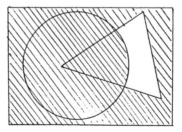

$$P(A \cup B^c) = P(A) + P(B^c) - P(AB^c)$$
$$= .5 + .7 - .29 = .91$$

1-R6 Wherever Mr. A sits, 2 of the remaining five sets are next to him; there are 2 chances in 5 that Mrs. A will occupy one of them.

1-R7 Along with p^4 there must be 8 factors q, and a coefficient which is $\binom{12}{4}$.

1-R8 (a) $(15)_4 = 15 \cdot 14 \cdot 13 \cdot 12 = 32{,}760.$

(b) $\binom{15}{5} = 3003.$

(c) Of $\binom{15}{4}$ equally likely combinations, $\binom{12}{4}$ are all white: $\binom{12}{4} / \binom{15}{4}$.

(d) This is the same as the probability that the first is black: 3/15.

(e) This is always equal to the ratio of the sample size to the population size. The calculation in this case (which can clearly be generalized) is

$$\frac{1 \cdot \binom{14}{3}}{\binom{15}{4}} = \frac{14!}{3!11!} \cdot \frac{4!11!}{15!} = \frac{4}{15}.$$

1-R9 (a) You can only choose the other 2 from the nonred ones: $\binom{7}{2} / \binom{8}{3}$.

(b) They are either all white or red: $\binom{3}{3} / \binom{8}{3} = 1/56.$

(c) This is the complement of the event in (b): 55/56.

(d) This is the same as the probability that the first is white: 2/8.

(e) $\frac{1}{8} \times \frac{2}{7} \times \frac{5}{6}.$

(f) Besides the 1 red (no choice here), you must pick 1 white from 2 and 1 blue from 5: $2 \times 5/\binom{8}{3} = 10/56.$

1-R10 $P(A^c B^c) = P[(A \cup B)^c] = 1 - P(A \cup B) = 1 - (.3 + .5 - 0) = .2.$

1-R11 (a) Your 5 must match exactly the one combination of 5 that wins *and* your powerball must be the one that wins: $\dfrac{1}{\binom{45}{5}} \times \dfrac{1}{\binom{45}{1}} = \dfrac{1}{54{,}979{,}155}.$

(b) You must choose 3 of the 5 winning numbers, choose 2 from the remaining 40 numbers, and choose the correct 1 powerball (out of 45):

$\dfrac{\binom{5}{3}\binom{40}{2}}{\binom{45}{5}} \times \dfrac{1}{45} \doteq .00014.$ [Side note: a \$1 ticket pays off only \$5!]

(c) Your 5 come from the 40 nonwinners and your power ball comes from the 44 nonwinners: $\dfrac{\binom{40}{5}}{\binom{45}{5}} \times \dfrac{44}{45} = .5266.$

1-R12 $P[E \cup (F \cup G)] = P(E) + P(F \cup G) - P[E(F \cup G)]$

$$= P(E) + [P(F) + P(G) - P(FG)] - P(EF \cup EG)$$
$$= P(E) + P(F) + P(G) - P(FG)$$
$$- [P(EF) + P(EG) - P(EFG)].$$

1-R13 In both cases $P(k) \geq 0$, so you need only check that $\sum P(k) = 1$:

(a) $\dfrac{1}{32} \displaystyle\sum_0^5 \binom{5}{k} = \displaystyle\sum_0^5 \binom{5}{k} \left(\dfrac{1}{2}\right)^k \left(\dfrac{1}{2}\right)^{5-k} = \left\{\dfrac{1}{2} + \dfrac{1}{2}\right\}^5 = 1$ (binomial theorem).

(b) Same idea as in (a): $\displaystyle\sum_0^5 \binom{5}{k} \left(\dfrac{1}{3}\right)^k \left(\dfrac{2}{3}\right)^{5-k} = \left\{\dfrac{1}{3} + \dfrac{2}{3}\right\}^5$.

CHAPTER 2

2-R1 (a) Sum the rows for marginal probabilities for Y and the columns for X. In both cases the possible values are 1, 2, 3, and probabilities .2, .4, .4.

(b) $P[(1, 2), (2, 3), (2, 1),$ or $(3, 2)] = 0 + .2 + 0 + .2 = .4.$

(c) Divide the probabilities in the $X = 3$ row by the row total: $\frac{1}{4}, \frac{2}{4}, \frac{1}{4}.$

(d) They are—their distributions are the same.

(e) No—the 0's are the easiest clue; the marginal probabilities are not 0 and cannot multiply to give a joint probability of 0.

(f) $f(2) = P[(1, 1)] = .1,$ $f(3) = P[(1, 2)$ or $(2, 1)] = 0,$

$f(4) = P[(1, 3), (2, 2),$ or $(3, 1)] = .4,$ $f(5) = P[(2, 3)$ or $(3, 2)] = .4.$

2-R2 (a) $P(AB) = 0,$ so $P(A \cup B) = P(A) + P(B) - 0 = .9.$

(b) $P(AB) = P(A)P(B) = .4 \times .5 = .2,$ so $P(A \cup B) = .4 + .5 - .2 = .7.$

2-R3 (a) $P(\text{male \& smoker}) = 12/50.$

(b) $P(\text{female}) = \dfrac{14 + 4 + 2}{50}.$

(c) Among 30 males, 12 are smokers: $12/30.$

(d) Among 26 smokers, 14 are female: $14/26.$

2-R4 $P(D) = P(D \mid MF) \cdot P(MF) + P(D \mid TWTh) \cdot P(TWTh)$

$$= .05 \times \frac{2}{5} + .01 \times \frac{3}{5} = \frac{.10}{5} + \frac{.03}{5}.$$

The terms in this expression define the odds, 10:3 and

$$P(MF \mid D) = \frac{P(MF \text{ and } D)}{P(D)} = \frac{.10/5}{.13/5} = \frac{10}{13}.$$

The tree diagram can help to understand the calculation.

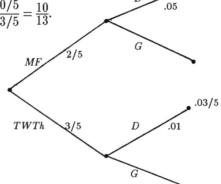

2-R5 (a) $P(X = 0) = P(Y = 0) = 1/2$ (row & column sums for $X = 0$, $Y = 0$). The events are independent: $P(X = 0 \text{ and } Y = 0) = 1/4 = (1/2) \times (1/2)$.

(b) The variables are *not* independent: the factorization that does hold for the events in (a), but does *not* hold in general. (The 0's in the table of probabilities are a give-away.) Neither are they exchangeable—the marginal p.f.'s are different.

2-R6 $1 = P(A \cup B \cup C)$
$\quad = P(A) + P(B) + P(C) - P(AB) - P(AC) - P(BC) + P(ABC)$
$\quad = .3 + .5 + .6 - .3 \times .5 - .3 \times .6 - .5 \times .6 + P(ABC) = .77 + P(ABC)$.
[We use the pairwise independence to factor: $P(AB) = P(A)P(B)$, etc]
This would mean $P(ABC) = .23$, which is larger than $P(AB) = .15$; and this can't be, because $ABC \subset AB$.

2-R7 This is easiest with a picture; label EF with its probability and deduce $P(E^c F)$ as $.3 - .2$. Also, because $P(E \cup F) = .6 = P(E) + .3 - .2$, it follows that $P(E) = .5 = P(E^c)$.

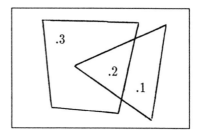

(a) Then, $P(F \cup E^c) = .3 + .5 - .1$.

(b) $P(F^c \mid E) = \dfrac{P(F^c E)}{P(E)} = \dfrac{.3}{.5}$.

2-R8 $f_{X,Y}(k, m) = f_X(k) \cdot f_Y(m) = 2^{-k} \times 2^{-m} = 2^{-(k+m)}$, $k, m = 1, 2, \ldots$.

2-R9 $P[E(F \cup G)] = P[EF \cup EG] = P(EF) + P(EG) - P(EFG)$
$\qquad\qquad\quad = P(E)P(F) + P(E)P(G) - P(E)P(FG)$
$\qquad\qquad\quad = P(E)[P(F) + P(G) - P(FG)] = P(E)P(F \cup G)$.

2-R10 (a) $P(\text{1st W} \mid \text{3rd W}) = P(\text{2nd W} \mid \text{1st W}) = 1/7$.

(b) $P(\text{2nd W} \mid \text{1st B \& 3rd R}) = P(\text{3rd W} \mid \text{1st B \& 2nd R}) = 2/6$.

2-R11 (a) Draws are independent, so probability for the 2nd is independent of the condition: $P(\text{2nd B} \mid \text{4th \& 5th W}) = P(\text{2nd B}) = P(\text{1st B}) = 1/3$.

(b) $P(\text{2nd B} \mid \text{4th \& 5th W}) = P(\text{3rd B} \mid \text{1st W \& 2nd W}) = 4/10$.

2-R12 $P(W) = P(W \mid A) \cdot P(A) + P(W \mid B) \cdot P(B)$
$= \frac{3}{4} \times .5 + \frac{1}{4} \times .5 = \frac{1.5}{4} + \frac{.5}{4}, \quad P(A \mid W) = \frac{P(W \mid A) \cdot P(A)}{P(W)} = \frac{1.5}{1.5 + .5} = \frac{3}{4}.$

2-R13 Let $q = 1 - p$: $P(\text{UDUUD}) = p \cdot q \cdot p \cdot p \cdot q = p^3 q^2.$

2-R14 (a) $P(X + Y = k \mid Y = j)$
$$= \frac{P(X + Y = k, \, Y = j)}{P(Y = j)} = \frac{P(X = k - j, \, Y = j)}{P(Y = j)}$$
Because of the assumed independence, the last numerator factors into the product of $P(X = k - j)$ and $P(Y = j)$; the latter cancels out.

(b) $P(X = j \mid X + Y = j + k) = \dfrac{P(X = j, X + Y = j + k)}{P(X + Y = j + k)}.$

Much as in (a), the numerator can be written $P(X = j, \, Y = k)$, which factors into the product $P(X = j) P(Y = k)$.

2-R15 (a) $f(0) = .3 + .2 + .1 = .6$, $f(1) = .2 + .1 = .3$, $f(2) = .1$. (Sum the probabilities in the rows and columns; it doesn't matter which is f_X and which is f_Y.)

(b) Divide the probabilities in the row (or column, as the case may be) for $X = 0$ by their total: $f_Y(y \mid x) = f(x, y)/f_X(0)$. The values of Y are 0, 1, 2, and their conditional probabilities are .3/.6, .2/.6, .1/.6, resp.

(c) $f_S(0) = f(0, 0) = .3$, $f_S(1) = f(1, 0) + f(0, 1) = .3$,
$f_S(2) = f(2, 0) + f(1, 1) + f(0, 2) = .4$, where $S = X + Y$.

(d) Not independent—see all the 0's; but they are exchangeable: $f(x, y)$ is symmetric in its arguments.

CHAPTER 3

3-R1 (a) $E2^X = \sum_0^2 2^k f(k) = 2^0 \times .6 + 2^1 \times .1 + 2^2 \times .3.$

(b) Same idea as (a): $Et^X = \sum_0^2 t^k f(k) = t^0 \times .6 + t^1 \times .1 + t^2 \times .3.$

(c) $X(X-1) = 0$ when $X = 0$ or 1; so $E[X(X-1)] = 2 \cdot 1 \cdot f(2) = .6.$

(d) $\text{var}(2X) = 4 \text{var} X = 4[E(X^2) - \mu_X^2] = 4[1.3 - .7^2] = 3.24.$

3-R2 $P(\text{odd} \mid \text{even}) = 0,\ P(2 \mid \text{even}) = \dfrac{1/6}{3/6} = \dfrac{1}{3} = P(4 \mid \text{even}) = P(6 \mid \text{even}).$

So, $E(X \mid \text{even}) = 2 \times \dfrac{1}{3} + 4 \times \dfrac{1}{3} + 6 \times \dfrac{1}{3} = 4.$

3-R3 (a) The marginal distributions are as follows:

x	$f_X(x)$	$x^k f_X(x)$	y	$f_Y(y)$	$y^k f_Y(y)$	
0	3/4	0	0	1/3	0	$EX^2 = 1/4,\ EY^2 = 2/3$
1	1/4	1/4	1	2/3	2/3	$\text{var} X = \frac{1}{4} - \left(\frac{1}{4}\right)^2 = \frac{3}{16}$
	$EX^k = 1/4 = EX$			$EY^k = 2/3 = EY$		$\text{var} Y = \frac{2}{3} - \left(\frac{2}{3}\right)^2 = \frac{2}{9}.$

(b) $f_{Y\mid 1}(0) = \dfrac{1/12}{1/4} = 1/3,\ f_{Y\mid 1}(1) = \dfrac{1/6}{1/4} = 2/3$: $E(Y \mid X = 1) = 2/3.$

(c) If you happen to notice that each joint probability is the product of the corresponding marginal probabilities—which means independence, you'd know that the covariance is 0. If not:

$$\text{cov}(X, Y) = \sum xy\, f(x, y) - \mu_X \mu_Y = \frac{1}{6} - \frac{1}{4} \cdot \frac{2}{3} = 0.$$

(d) $\text{cov}(X - 2Y, 3X + Y)$

$$= \text{cov}(X, 3X) + \text{cov}(X, Y) - \text{cov}(2Y, 3X) - \text{cov}(2Y, Y)$$

$$= 3\,\text{cov}(X, X) + \text{cov}(X, Y) - 6\,\text{cov}(X, Y) - 2\,\text{cov}(Y, Y)$$

$$= 3\,\sigma_X^2 + \sigma_{X,Y} - 6\sigma_{X,Y} - 2\sigma_Y^2 = 3 \cdot \frac{3}{16} - 0 - 2 \cdot \frac{2}{9}.$$

3-R4 (a) $\sigma_{X,X} - \sigma_{X,Y} + \sigma_{Y,X} - \sigma_{Y,Y} = \text{var } X - \text{var } Y = 5.$

(b) $\text{var } X + 4\,\text{cov}(X, Y) + 4\,\text{var } Y = 9 - 20 + 16.$

(c) $\text{cov}(X, X) + 2\,\text{cov}(X, Y) = \sigma_X^2 + 2\sigma_{X,Y} = 9 + 2 \times (-5) = -1.$

(d) $\rho_{X, X+2Y} = \dfrac{\text{cov}(X, X+2Y)}{\sqrt{\text{var } X}\sqrt{\text{var}(X+2Y)}} = \dfrac{-1}{3 \times \sqrt{5}}.$

3-R5 (a) $f(k)$ is the coefficient of t^k: $f(0) = .3$, $f(1) = .2$, $f(2) = .5.$

(b) $EX = 1 \times .2 + 2 \times .5 = 1.2$, and $E(X^2) = 1 \times .2 + 2^2 \times .5 = 2.2.$

Then, $\text{var } X = 2.2 - 1.2^2 = .76.$

3-R6 The mean is $5/2$ (symmetry); $E(X^2) = \frac{1}{4}(1 + 4 + 9 + 16) = \frac{15}{2}$, so

$\sigma^2 = E(X^2) - (EX)^2 = \frac{15}{2} - \frac{25}{4}.$

3-R7 $\text{cov}(X + Y, X - Y) = \text{var } X + \text{cov}(Y, X) - \text{cov}(X, Y) - \text{var } Y = 0,$
since X and Y have the same distribution and hence the same variance.

3-R8 (a) Only one nonzero product has positive probability: $E(XY) = 1 \times .1.$

(b) For this we need variances. The two marginal variables are identically distributed, with

$$\mu = 1 \times .3 + 2 \times .1 = .5, \text{ and } E(X^2) = 1 \times .3 + 4 \times .1 = .7,$$

so $\text{var } X = \text{var } Y = .7 - .25 = .45.$ Then,

$$\sigma_{X,Y} = E(XY) - \mu_X\mu_Y) = .1 - .5 \times .5 = -.15,$$

and

$$\rho = \frac{\sigma_{X,Y}}{\sigma_X\sigma_Y} = \frac{-.15}{\sqrt{.45}\sqrt{.45}} = \frac{-.15}{.45} = -\frac{1}{3}.$$

(c) The conditional p.f. of X given $Y = 0$ is found by dividing the entries in the column for $Y = 0$ by their total, to get $3/6$, $2/6$, $1/6$. The mean is $E(X \mid Y = 0) = 0 \times 3/6 + 1 \times 2/6 + 2 \times 1/6 = 2/3.$

(d) This is $EX + EY = .5 + .5.$

(e) $\text{var}(X + Y) = \text{var } X + \text{var } Y + 2\,\text{cov}(X, Y) = 2 \times .45 + 2 \times (-.15).$

3-R9 (a) The coefficient of t^k is $P(X = k) = f(k)$, and

$$EX = \tfrac{1}{81}(1 \times 32 + 2 \times 24 + 3 \times 8 + 4 \times 1) = 108/81 = 4/3.$$

(b) $\eta'(t) = \frac{1}{81}(32 + 48t + 24t^2 + 4t^3).$ Substituting $t = 1$ produces (a).

3-R10 (a) There are 10 possible samples, equally likely:

$$012, \ 013, \ 014, \ 023, \ 024, \ 034, \ 123, \ 124, \ 134, \ 234.$$

Now just count: $P(0, 3) = P[(013 \text{ or } 023)] = 2/10$, and so on:

Y:

X:	2	3	4	
0	.1	.2	.3	.6
1	0	.1	.2	.3
2	0	0	.1	.1
	.1	.3	.6	1

(b) $EX = 1 \times .3 + 2 \times .1 = .5$, $E(XY) = 3 \times .1 + 4 \times .2 + 8 \times .1 = 1.9$,

$EY = 2 \times .1 + 3 \times .3 + 4 \times .6 = 3.5$, so then $\sigma_{X,Y} = 1.9 - .5 \times 3.5$.

(c) $f_{X\,|\,4}(1) = 2/6$, $f_{X\,|\,4}(2) = 1/6$: $E(X\,|\,4) = 4/6$, $E(X^2\,|\,4) = 6/6$.
Then, var$(X\,|\,4) = E(X^2\,|\,4) - [E(X\,|\,4)]^2 = 1 - (4/6)^2 = 5/9$.

3-R11 $P(E) = \sum_x P(E\,|\,X = x)f(x) = \sum g(x)f(x) = E[g(X)]$.

3-R12 (a) The distribution for each of the marginal variables is the same, as shown below in the margins.

	0	1	2	3	
0	.04	.08	.06	.02	.2
1	.08	.16	.12	.04	.4
2	.06	.12	.09	.03	.3
3	.02	.04	.03	.01	.1
	.2	.4	.3	.1	

To represent independence, we find the joint probabilities by multiplying corresponding marginal probabilities. For the sum, add probabilities along diagonals—for instance,

$$P(X + Y = 2) = P[(2, 0), \ (1, 1), \ (0, 2)] = .06 + .16 + .06 = .28.$$

(b) $.04 + .16t + .28t^2 + .28t^3 + .17t^4 + .06t^5 + .01t^6$.

(c) Square the p.g.f. for X, because the variables are independent:

$$(.2 + .4t + .3t^2 + .1t^3)^2.$$

To square the polynomial, square each term, take twice each cross-product, and add:

$$.04 + (2 \times .08)t + (.16 + 2 \times .06)t^2 + (2 \times .02 + 2 \times .12)t^3$$
$$+ (.09 + 2 \times .04)t^4 + (2 \times .03)t^5 + .01t^6$$

CHAPTER 4

4-R1 $p = .25$, $n = 300$, $EY = np = 75$, var $Y = npq = 56.25 = 7.5^2$. Since npq is large, we can use a normal approximation:

$$P(Y \leq 67) \doteq \Phi\left(\frac{67 - 75}{7.5}\right) = .143.$$

With a continuity correction,

$$P(Y \leq 67) \doteq \Phi\left(\frac{67.5 - 75}{7.5}\right) = .1587.$$

4-R2   (a) The number of trials for a first success is Geo(.04), with mean $1/.04$, or 25. To get six successes, it takes 6×25 trials, on average.

(b) $X_i \sim$ Bin(50, .04) with mean $50 \times .04 = 2$ and variance $2 \times .96 = 1.92$. So, $E(X_1 + X_2) = EX_1 + EX_2 = 4$, var$(X_1 + X_2) = 2$ var $X_1 = 3.84$.

(c) The sum of these independent binomials is binomial with the same p and $n = n_1 + n_2 = 100$: $Y \sim$ Bin(100, .04).

(d) The n is large and the p small; $np = 4$, so $Y \sim$ Poi(4):

$$P(Y \leq 5) = \sum_0^5 \frac{4^k}{k!} e^{-4}.$$

Table IV gives this as .785 ($m = 4$, $c = 5$).

(e) $P(X_1 = 3, X_2 = 3 \mid Y = 6) = \dfrac{P(X_1 = 3)P(X_2 = 3)}{P(Y = 6)}$

$$= \frac{\binom{50}{3}.04^3.96^{47} \times \binom{50}{3}.04^3.96^{47}}{\binom{100}{6}.04^6.96^{94}} = .32227,$$

or (using Poisson approximations)

$$\doteq \frac{(e^{-2}2^3/3!)(e^{-2}2^3/3!)}{e^{-4}4^6/6!} = \binom{6}{3}\frac{1}{2^6} = .3125.$$

4-R3   (a) Let $Y = \#$(males). $EY = 5 \times 10/25 = 2$, and $P(Y = 2) = \dfrac{\binom{10}{2}\binom{15}{3}}{\binom{25}{5}}$.

(b) $P(Y \geq 3) = P(Y = 3, 4, \text{ or } 5) = \dfrac{\binom{10}{3}\binom{15}{2} + \binom{10}{4}\binom{15}{1} + \binom{10}{5}\binom{15}{0}}{\binom{25}{5}} \doteq .301.$

4-R4 (a) According to the table at the end of §4.9, the p.g.f. of Geo(p) is $pt(1 - qt)^{-1}$. The p.g.f. of the sum of 4 independent observations on any variable is the 4th power of the p.g.f. for 1: $[pt(1 - qt)^{-1}]^4$.

(b) The mean of Geo(p) is $1/p$, so $EY = 4EX = 4/p$. The variance of Geo(p) is q/p^2, so $\text{var } Y = 4q/p^2$.

(c) There must be 3 successes in the first 6 trials, followed by another success: $\binom{6}{3}p^3q^3 \cdot p = \binom{6}{3}p^4q^3$.

4-R5 (a) With $\lambda = .5$ per page the number in 20 pages is Poi($20 \times .5$), and the mean and the variance are both 10.

(b) In 6 pages, the mean is $6\lambda = 3$; enter Table IV with $m = 3$, $c = 3$:

$$.647 = e^{-3}(1 + 3 + 3^2/2! + 3^3/3!).$$

4-R6 In a single toss of 2 coins, you get 2 heads with probability 1/4, 2 tails with probability 1/4, and one of each with probability 1/2. Then, in a a random sample, the distribution of the frequencies of 0, 1, and 2 is multinomial, with $n = 9$ and corresponding probabilities 1/4, 2/4, 1/4:

$$f(3, 3, 3) = \frac{9!}{3!3!3!}\left(\frac{1}{4}\right)^3\left(\frac{2}{4}\right)^3\left(\frac{1}{4}\right)^3 = .05127.$$

4-R7 (a) Another multinomial situation: $f(4, 4, 2) = \binom{10}{4,\,4,\,2}.35^4.35^4.30^2$.

(b) $G \sim \text{Bin}(10, .35)$, with mean $np = 3.5$.

(c) $G + R = 10 - B \sim \text{Bin}(10, .35 + .35)$.

4-R8 (a) Each selection is a Ber(1/4), and we assume independence of the 40 trials, so the number correct is Bin($40, .25$), and the expected number correct is $40 \times 1/4$.

(b) Since $npq = 7.5 > 5$, we use a normal approximation:

$$P(\text{at most } 8) \doteq \Phi\left(\frac{8.5 - 10}{\sqrt{7.5}}\right) = \Phi(-.548) = .292.$$

4-R9 (a) $\lambda = 3/\text{min}$, so for 2 minutes, $\lambda t = 6$:

$$P(\text{at least } 5) = 1 - P(\text{at most } 4) = 1 - .285.$$

(from Table IV, with $\mu = 6$, $c = 4$).

(b) $T = 20$ seconds $= 1/3$ minute, so $\lambda T = 1$: $P(2) = e^{-1}1^2/2! = .184$.

(c) The numbers in the first 10-second (1/6 min) period and the second 10-second period are independent, each Poisson, with $\lambda T = 3 \times 1/6 = 1/2$. The number in 20 seconds [see (b)] is Poi(1). Then (by the definition of conditional probability)

$$P[1 \text{ in } (0, 10), 1 \text{ in } (10, 20) \mid 2 \text{ in } (0, 20)] = \frac{\frac{1}{2}e^{-1/2} \times \frac{1}{2}e^{-1/2}}{e^{-1}1^2/2\,!} = \frac{1}{2}.$$

(d) The future is independent of the past, so the condition "eight calls in the preceding 5-minute period" is irrelevant. With $T = 4$, $\lambda T = 12$, and $P(\text{at most } 6) = .046$, from Table IV ($m = 4$, $c = 6$).

4-R10 (a) Because $N \gg n$, the number of Republicans in a sample of size 5 is approximately Bin(5, .3): $f(2) = \binom{5}{2}.3^2.7^3$.

(b) The population is "large," so we can approximate the distribution of the number in a random sample of 200 using Bin(200, .04). Since n is large and p small, we can approximate this using Poi($200 \times .04$):

$P(\text{at least } 10) = 1 - P(\text{at most } 9) = 1 - .717$ (Table IV: $m = 8$, $c = 9$).

[This is apt to be closer to the exact value than what we get using the binomial approximation: $1 - \Phi\left(\dfrac{9.5 - 8}{\sqrt{8 \times .04 \times .96}}\right) = .294.$]

4-R11 We use Poisson, with $m = np = 55{,}000{,}000 \times \dfrac{1}{54{,}979{,}155} \doteq 1.00379$.

(a) $P(0 \text{ winners}) = e^{-1.00379} \doteq .3665.$ (Rounding to $m = 1$ gives .368.)

$P(1 \text{ winner}) \doteq e^{-1} \times 1/1! = .368$, and $P(\text{at least } 1) = 1 - P(\text{none}) = .632.$

(b) If $n = 20{,}000{,}000$, $np = .36377$: $P(0) = e^{-.36377} = .695$,

$P(\text{exactly } 1) = \dfrac{.36377}{1!}e^{-.36377} = .695$, and $P(\text{at least } 1) = 1 - P(\text{none}).$

CHAPTER 5

5-R1 (a) $P(X = 1)$ is the size of the jump at 1: $F(1) - F(1-) = 1/2 - 0.$

(b) There is no jump at $x = 2$.

(c) $P(1 < X < 2) = F(2) - F(1) = 3/4 - 1/2.$

5-R2 (a) It takes two formulas to define f, so it takes two to define the c.d.f.:
For $0 < y < 1$,
$$F(y) = \int_0^y \frac{2u}{3} \, du = \frac{y^2}{3}.$$

(This is just the area of a triangle
with base y and altitude $2y/3$—
it's not necessary to integrate,
although integration works.)

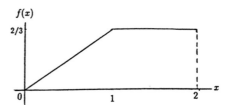

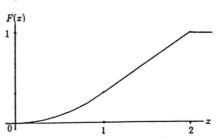

When $1 < y < 2$, the accumulated area up to y is $1/3$ (the amount to the left of 1) *plus* the area from 1 to y: $1/3 + (y - 1) \times 2/3$. [The latter area is that of a rectangle of height $f(y) = 2/3$ and base $y - 1$.]

(b) The area between 0 and 1 is $1/3$, so the first quartile is less than 1, where $F(y) = y^2/3$. Setting this equal to $1/4$, we solve for y: $y = \sqrt{3/4}$.

(c) The median is to the right of 1, where $F(y) = (2y - 1)/3$, which equals $1/2$ at $y = 5/4$.

(d) $EY = \displaystyle\int_0^1 y \cdot \frac{2y}{3} \, dy + \int_1^2 y \cdot \frac{2}{3} \, dy = \frac{2}{9} + 1 = \frac{11}{9}.$

(e) $E(Y^2) = \displaystyle\int_0^1 y^2 \cdot \frac{2y}{3} \, dy + \int_1^2 y^2 \cdot \frac{2}{3} \, dy = \frac{93}{54}$, var $Y = \frac{93}{54} - \left(\frac{11}{9}\right)^2 = \frac{37}{162}.$

5-R3 (a) Since $[V < 2] \subset [V < 1]$, $P(V < 1 \mid V < 2) = \dfrac{P(V < 1)}{P(V < 2)} = \dfrac{1/3}{2/3}$.

(b) $F'(v) = 1/3$, $0 < v < 3$. $[V \sim \mathcal{U}(0, 3).]$

(c) Dividing each v between 0 and 3 by 3 yields $\mathcal{U}(0, 1)$. Formally, we have $P(V/3 < u) = P(V < 3u) = F_V(3u) = u$, for $0 < u < 1$.

5-R4 $F_Y(y) = P(Y \le y) = P(\sqrt{X} \le y) = P(X \le y^2) = F_X(y^2) = y^2$, on $(0, 1)$.

5-R5 Integrating f we find $F(x) = x^3$, $0 < x < 1$. If $y = -\log x$, $x = e^{-y}$, and

$$f_Y(y) = f_X(e^{-y})\left|\frac{de^{-y}}{dy}\right| = 3e^{-3y}, \; y > 0.$$

[See (15) in §5.2, page 177 of the text.]

5-R6 (a) Use symmetry: $f(-u) = f(u)$.

(b) $\mu_U = 0$, so $\operatorname{var} U = E(U^2) = \displaystyle\int_{-1}^{1} u^2 \cdot \frac{3}{10}(2 - u^2)\, du$. The integrand is symmetric about 0, so we can integrate from 0 to 1 and multiply by 2:

$$\operatorname{var} U = \frac{3}{5}\int_0^1 (2u^2 - u^4)\,du = \frac{2}{5} - \frac{3}{25} = \frac{7}{25} = (.529)^2.$$

5-R7 (a) $E(e^{-X}) = \psi_X(-1) = (1 + 2)^{-1/2}$.

(b) $\psi'(t) = -\frac{1}{2}(1 - 2t)^{-3/2} \cdot (-2)$, and $EX = \psi'(0) = 1$. Similarly,

$\psi''(t) = -\frac{3}{2}(1 - 2t)^{-5/2} \cdot (-2)$, and $E(X^2) = \psi''(0) = 3$: $\operatorname{var} X = 3 - 1^2$.

5-R8 (a) The region where $X < Y$ is half the circle, and probability $\propto$ area.

(b) When the distribution is uniform, the joint p.d.f. is

$$\frac{1}{\text{area of the support}} = \frac{1}{\pi}.$$

We get the marginal p.d.f. of X by integrating over y (across the chord of the circle at x):

$$f_X(x) = \int_{-a}^{a} \frac{1}{\pi}\,dy = \frac{2a}{\pi}, \text{ where } a = \sqrt{1 - x^2}.$$

(c) Symmetry around 0 implies zero covariance.

5-R9 (a) According to the definition of independent variables,

$$f_{U,V}(u, v) = f_U(u) \cdot f_V(v) = ue^{-u} \cdot ve^{-v} \text{ for } u > 0 \text{ and } v > 0.$$

(b) Independence implies $\rho = 0$.

(c) $\psi_X(t) = E(e^{tX}) = \displaystyle\int_0^\infty e^{tx} \cdot xe^{-x}dx = \int_0^\infty xe^{-(1-t)x}dx = \dfrac{1}{(1-t)^2}.$

Then, independence implies $\psi_{U+V}(t) = \psi_U(t) \cdot \psi_V(t) = \dfrac{1}{(1-t)^4}.$

5-R10 (a) The joint p.d.f. is 2 on the support—a triangle whose area is $1/2$.

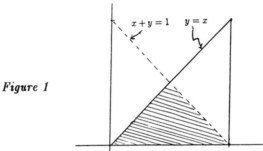

Figure 1

To get the marginal p.d.f. we integrate out the y:

$$f_X(x) = \int_0^x 2\,dy = 2x, \ 0 < x < 1.$$

(b) When the joint p.d.f. is constant, conditional p.d.f.'s are constant. Thus, $Y \mid x$ is uniform on the interval $0 < y < x$, so its mean is $x/2$.

c. The region below the line $x + y = 1$ (Fig. 1) is half the support region.

(d) $E(XY) = \displaystyle\int_0^1\int_0^x 2xy\,dy\,dx = \int_0^1 2x \cdot \dfrac{x^2}{2}\,dx = \dfrac{1}{4}.$

5-R11 (a) Symmetry about the y-axis (Figure 2) implies $EX = 0$.

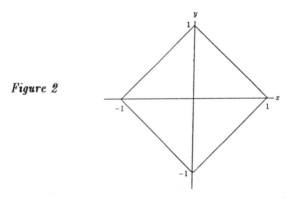

Figure 2

(b) The area of the support is 2, so the p.d.f. is the constant 1/2. Then,

$$f_X(\tfrac{1}{2}) = \int_{-1/2}^{1/2} \tfrac{1}{2}\,dy = \tfrac{1}{2}.$$

(c) $f_{Y\,|\,\frac{1}{2}}(y) = \dfrac{f(\frac{1}{2},\,y)}{f_X(\frac{1}{2})} = \dfrac{1/2}{1/2}.$

(d) Symmetry about the origin implies zero covariance.

(e) X and Y are exchangeable, with common variance $\operatorname{var} X = E(X^2)$ (since $EX = 0$). In integrating to find this expected value, we can integrate over the first quadrant and multiply by 4, owing to the symmetry with respect to each axis:

$$E(X^2) = 4 \times \tfrac{1}{2}\int_0^1 x^2 \int_0^{1-x} dy\,dx = 2\int_0^1 x^2(1-x)\,dx = \tfrac{1}{6}.$$

Now we can do what's called for:

$$\operatorname{var}(X + 2Y) = \operatorname{var} X + 4\operatorname{var} Y = 5\operatorname{var} X = 5/6.$$

5-R12 (a) The joint p.d.f. (including the support) factors into the product of a function of x and a function of y: $f(x,\,y) = xe^{-x^2/2}\cdot ye^{-y^2/2}$, $x,\,y > 0$.

(b) $F_X(x) = \displaystyle\int_0^x ue^{-u^2/2}\,du = 1 - e^{-x^2/2}$, $x > 0$. [We integrated by parts, with $u^2/2 = v$, $dv = u\,du$; the resulting v-integral has limits 0 and $x^2/2$.]

(c) $E(Y^2) = E(X^2) = \displaystyle\int_0^\infty x^2\left(e^{-x^2/2}\right)x\,dx$, and again set $x^2/2 = v$:

$$E(X^2) = \int_0^\infty 2ve^{-v}\,dv = 2.$$

5-R13 $P(|X| < .02) = P(-.02 < X < .02) = 2P(0 < X < .02) \doteq 2f(0) \times .02.$

The p.d.f. is $f(x) = \dfrac{\pi}{\sqrt{2}}e^{-\pi x/\sqrt{2}} \times \left(1 + e^{-\pi x/\sqrt{2}}\right)^{-2}$, so $f(0) = \dfrac{\pi}{4\sqrt{2}}.$

The approximate probability is $2f(0) \times .02 = .0222.$

5-R14 (a) The constant density is the reciprocal of the content of the region of support, in this case the volume of the tetrahedron: $3 \times$ area of base $= 6$.

(b) When a p.d.f. is constant, conditional p.d.f.'s are constant. Given $X = x$ and $Y = y$, the variable Z can vary from 0 up to the value of Z on the plane $x + y + z + 1$. So Z has constant density on $(0, 1 - x - y)$.

(c) The variables are exchangeable. To find the p.d.f. of X, we need to integrate over y and z. For $0 < x < 1$, the integral is

$$\int_0^{1-x} \int_0^{1-x-z} 6 \, dy \, dz = \int_0^{1-x} 6(1 - x - z) \, dz = 3(1 - x)^2.$$

(The inner integral sums the volume elements from the xy-plane up to the plane $z = 1 - x - y$; the outer integral sums these columns from $y = 0$ out to the line $x + y = 1$. See Figure 3.)

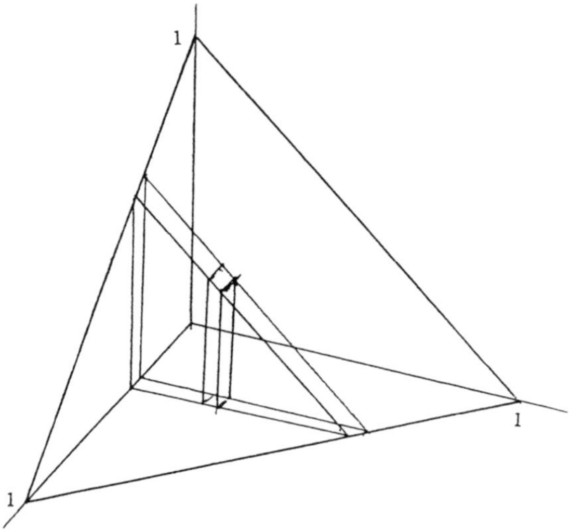

Figure 3

5-R15 (a) The mean is the coefficient of t: $\mu = 1$, and the average square is the coefficient of $t^2/2$: $E(X^2) = 2$, so var $X = 2 - 1^2 = 1$.

(b) $\text{var}(3x - 2) = 3^2 \text{var } X = 9$.

5-R16 (a) F is continuous at 2, so $P(X = 2) = 0$.

(b) F jumps from 0 to .1 at $x = 0$, so $P(X = 0) = .1$.

(a) $P(X > 2) = 1 - P(X \le 2) = 1 - F(2) = .9 \, e^{-2} = .122$.

(d) The median m is the number for which $F(m) = .5$, or $.9e^{-m} = .5$, or $-m = \log(5/9)$, or $m \doteq .588$.

5-R17 $P(A \mid Y = y) = \displaystyle\int_A f(x \mid y)\, dx = \int_A \frac{f(x,y)}{f_Y(y)}\, dx = g(y).$

$$Eg(Y) = \int_{-\infty}^{\infty} g(y) f_Y(y)\, dy = \int_{-\infty}^{\infty} \left\{ \int_A \frac{f(x,y)}{f_Y(y)}\, dx \right\} f_Y(y)\, dy$$

$$= \int_{-\infty}^{\infty} \left\{ \int_A f(x,y)\, dx \right\} dy = P(X \in A).$$

5-R18 (a) *Discrete case:*

$$P(X = x_i) = \sum P([X = x_i] \cap A_i) = \sum P(X = x_i \mid A_i) P(A_i).$$

Continuous case:

Applying the same idea to the event $X \le x$, we have

$$P(X \le x) = \sum P([X \le x] \cap A_i) = \sum P(X \le x \mid A_i) P(A_i),$$

or

$$F(x) = \sum_i F_{X \mid A_i}(x) P(A_i).$$

Differentiating yields the desired equality.

(b) *Discrete:* $EX = \displaystyle\sum x_j f(x_j) = \sum_j x_j \sum_i f_{X \mid A_i}(x_j) P(A_i)$

$$= \sum_i P(A_i) \sum_j x_j f_{X \mid A_i}(x_j) = \sum P(A_i) E(X \mid A_i).$$

Continuous: $EX = \displaystyle\int x f(x)\, dx = \int x \sum f_{X \mid A_i}(x) P(A_i)\, dx$

$$= \sum P(A_i) \int x f_{X \mid A_i}(x)\, dx = \sum P(A_i) E(X \mid A_i).$$

CHAPTER 6

6-R1 (a) Except for a constant factor, the integral is the area under the graph of the p.d.f. of $N(3, 1/2)$; the missing constant is $\dfrac{1}{\sqrt{2\pi\sigma^2}} = \dfrac{1}{\sqrt{\pi}}$.

(b) The integrand, except for a constant factor, is the p.d.f. of $\text{Chi}^2(9)$. The missing factor is $= [2^{9/2}\Gamma(\tfrac{9}{2})]^{-1}$, so the integral is

$$2^{9/2}\Gamma(\tfrac{9}{2}) \cdot F_{x^2(9)}(15) = 2^{9/2}[\tfrac{7}{2} \cdot \tfrac{5}{2} \cdot \tfrac{3}{2} \cdot \tfrac{1}{2}\sqrt{\pi}] \times .909 = 239.2.$$

6-R2 (a) The given p.d.f. is that of $N(2, 4)$, so $K = 1/\sqrt{8\pi}$.

(b) A linear function of a normal variable is normal; so Y is normal, with mean $\tfrac{1}{2}(EX - 2) = 0$, and s.d. $\tfrac{1}{2}\sigma_X = 1$.

(c) Substitute $t = 1$ in the m.g.f., $\psi_X(t) = \exp(2t + 4t^2/2\ )$: $\psi(1) = e^4$.

6-R3 (a) The sum of independent normal variables is normal; the mean is the sum of the means (in this case, $5n$), and the variance is the sum of the variances $(4n)$.

(b) The joint p.d.f. is the product of the marginals:

$$\prod_i f_{X_i}(x_i) = \prod (2\pi \times 4)^{-1/2}\exp[-(x_i - 5)^2/8].$$

(c) We take the covariance of each term in $2X_1 - X_2$ with each term in $X_1 + 2X_2$, factor out constant coefficients, and sum:

$$2\operatorname{cov}(X_1, X_1) + 4\operatorname{cov}(X_1, X_2) - \operatorname{cov}(X_2, X_1) - 2\operatorname{cov}(X_2, X_2)$$
$$= 2\sigma_1^2 + 3\sigma_{12} - 2\sigma_2^2 = 8 - 0 - 8 = 0.$$

6-R4 In Table IV, we find $P(0) = .135$ for $m = \lambda t = 2$, so $\lambda = 2/10 = .2/\text{min}$.

(a) The mean time to the first arrival after any point in time is $1/\lambda = 5$, so the waiting time to the sixth is 6×5 minutes.

(b) The information about arrivals in the preceding interval is irrelevant, because of independence. So the answer is $P(T < 5) = 1 - e^{-5 \times 1/5}$.

(c) The time exceeds 30 minutes if (and only if) in that period there are 5 or fewer arrivals. The mean number in 30 minutes is $30 \times .2 = 6$, so

$$P(0, 1, ..., \text{or } 5) = \sum_0^5 e^{-6}6^k/k! = .446 \text{ (Table IV)}.$$

6-R5 (a) The sum of independent, standard normals is normal: $Y \sim N(0, 3)$,
so $P(X_1 + X_2 + X_3 > 4) = 1 - P(\sum X_i < 4) = 1 - \Phi\left(\frac{4-0}{\sqrt{3}}\right) \doteq \Phi(-2.31)$.

(b) The sum of squares of 3 independent, standard normals is chi$^2(3)$, so
$P(X_1^2 + X_2^2 + X_3^2 > 4) = 1 - F_{\chi^2(3)}(4) = .261$ (Table Vb).

(c) Because of independence, the conditional distribution of X_1, given any condition on the other two variables, is just standard normal.

6-R6 (a) Compare the given m.g.f. with that of $N(\mu, \sigma^2)$: $\exp\{\mu t + \sigma^2 t^2 / 2\}$.
We see that $\mu = 2$, and $\sigma^2 / 2 = 2$, or $\sigma^2 = 4$.

(b) $P(X < 0) = \Phi\left(\frac{0-2}{2}\right) = \Phi(-1) = .1587$.

6-R7 This m.g.f. is the cube of $(1 - t/\lambda)^{-1}$, where $\lambda = 1/4$, which is the m.g.f. of Exp(1/4). The cube is the m.g.f. of the sum of three independent copies of Exp(1/4), and this defines Gam(3, 1/4).

6-R8 (a) $E(X^2) = \sigma^2 + \mu^2 = 25 + 4$.

(b) A linear transformation of X is *normal*: $EY = 2(EX + 2) = 14$, and var $Y = 2^2$ var $X = 4 \times 4$.

(c) $(X - 5)/2$ is the Z-score of X—its distribution is $N(0, 1)$; so its square is chi$^2(1)$.

6-R9 With the given waiting time distribution, this is a Poisson process with $\lambda = 1/3$ per week (the mean time to failure is $1/\lambda$).

(a) $P(T > 12) = P(0, 1, 2, \text{ or } 3 \text{ in } 12 \text{ wks.}) = \sum_0^3 e^{-4} 4^k / k! = .433$.
(This can be found in Table IV: $m = 12\lambda = 4$, $c = 3$).

(b) The future is independent of information about prior failures, so this is just exponential with $\lambda = 1/3$, and $P(T < 6) = F(6) = 1 - e^{-6/3}$.
(You can also think of this as 1 minus the probability of 0 failures in six weeks and use the Poisson formula.)

6-R10 (a) The joint p.d.f. is the product of the marginals:
$$\prod_i f_{X_i}(x_i) = e^{-x_1} e^{-x_2} e^{-x_3}, \text{ for all } x_i > 0.$$
(b) The sum is Gam(3, 1) (see §6.3).

CHAPTER 7

7-R1 $S^2 = \frac{1}{n-1}\sum(X_i - \bar{X})^2 = \frac{1}{n-1}\left\{\sum X_i^2 - \frac{(\sum X_i)^2}{n}\right\}$, so $\sum X_i^2 = 9S^2 + \frac{5^2}{10}$.

7-R2 $12 = \frac{1}{9}(198 - 10\bar{X}^2)$, $-10\bar{X}^2 = 108 - 198$, or $\bar{X}^2 = 9$.

7-R3 $\sum\sum(X_i - X_j)^2 = \sum\sum(X_i^2 - 2X_iX_j + X_j^2) = 2n\sum X_i^2 - 2(\sum X_i)^2$

$$= 2n\left\{\sum X_i^2 - (\sum X_i)^2/n\right\} = 2n\sum(X_i - \bar{X})^2 = 2n^2V.$$

7-R4 (b) When counting to the 10th and 11th smallest, remember to enter the first 2-stem at the 1's; the 10th is a 2 and the 11th is a 3.

The quartiles are 18.5 and 28, so IQR $= 9.5$.

(c) Use a statistical calculator if possible. Otherwise, for you to check: $\sum X_i = 458$, $\sum X_i^2 = 11{,}188$.

7-R5 (b) $\sum X_i = 306.4$, $\sum X_i^2 = 5934.72$, $\sum Y_i = 1272$, $\sum Y_i^2 = 89280$, and $\sum X_iY_i = 16{,}490.1$.

7-R6 (a) Dot diagrams (on the same axis—one above and one below) are about as good as any.

(c) 2.4, 3.4, 3.2, $-.3$, 0, $-.7$, $-.2$, 2.8, 3.9, 2.5, .1, $-.2$, -9, -1.6.

7-R7 (a) Either find $\bar{B}$ and $\bar{A}$ and subtract, or find $D_i = B_i - A_i$ and $\bar{D}$.

7-R8 (a) $S_U = S_X = 3$, $S_V = 2S_Y = 10$.

(b) With linear transformations, there is no change in r.

7-R9 The median and quartiles are a little easier to find if you first order the leaves on each stem. Otherwise, don't forget to order them mentally: counting in on the 13 stem we have 13.1 as the 19th smallest, 13.2 as the 20th, etc. The quartiles are 12.0 and 14.8.

7-R10 (a)

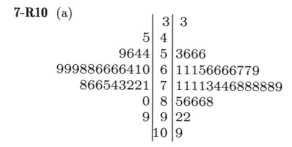

```
              |  3 | 3
            5 |  4 |
         9644 |  5 | 3666
 999886666410 |  6 | 11156666779
    866543221 |  7 | 11113446888889
            0 |  8 | 56668
            9 |  9 | 22
              | 10 | 9
```

(b)

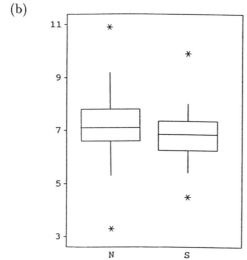

7-R11 $R = \frac{d}{dx}\sum|X_i - a| = \sum\frac{d|a - X_i|}{dx} = \sum[\text{sign of }(a - X_i)]$. Now think of starting to the left of all the data, where all signs of $a - X_i$ are negative; the derivative of R is $-n$. Next move to the right; as you pass each of the observations, one sign changes, and the sum increases by 2. If n is odd, the derivative goes from -1 to 1 as you pass the middle observation, If n is even, the derivative is 0 between the middle 2 obervations. In either case, the minimum value of the sum is at the median.

CHAPTER 8

8-R1 $L(p) = p^3(1-p)$: $\dfrac{L(\frac{1}{2})}{L(\frac{1}{3})} = \dfrac{(\frac{1}{2})^4}{(\frac{1}{3})^3(\frac{2}{3})} = \dfrac{3^4}{2^5}$.

8-R2 (a) $\prod f(x_i \mid \theta) = \prod \theta x_i^{\theta-1} = \theta^n (\prod x_i)^{\theta-1}$.

(b) The likelihood function is the function of θ in (a), for fixed x_i; its value is determined when you know θ and $\prod x_i$. So the product of the observations is sufficient for θ.

(c) $F(x) = \displaystyle\int_0^x \theta u^{\theta-1} du = x^\theta$, $0 < x < 1$. Then, with $Y = X_{(n)}$,

$$F_Y(y) = P(X_{(n)} \le y) = [F(y)]^n = y^{n\theta}, \ 0 < y < 1.$$

(d) $E[X_{(n)}] = \displaystyle\int_0^1 y \, f_Y(y) \, dy = \int_0^1 y[n\theta y^{n\theta-1}] dy = \dfrac{n\theta}{n\theta+1}$.

8-R3 (a) $L(\theta)$ when $X = 1$ is $(.5, .6, .3, .7)$ for $(\theta_1, \theta_2, \theta_3, \theta_4)$, respectively. Its maximum is at θ_4.

(b) For $k = 1, 3$: $P(X = k \mid X = 0 \text{ or } 2) = \dfrac{P(X = k \text{ and } X = 0 \text{ or } 2)}{P(X = 0 \text{ or } 2)} = 0$.

But

$$P(X = 0 \mid X = 0 \text{ or } 2) = \dfrac{P(X = 0 \text{ and } X = 0 \text{ or } 2)}{P(X = 0 \text{ or } 2)}$$

$$= \dfrac{P(X = 0)}{P(X = 0 \text{ or } 2)} = \dfrac{2}{3}.$$

Similarly, $P(X = 2 \mid X = 0 \text{ or } 2) = 1/3$.

(c) The statistic $|X - 1|$ defines the partition $\{0, 2\}, \{1\}, \{3\}$. Because the conditional probabilities in (b) are independent of the parameter θ, the partition is sufficient for θ, according to the general definition given at the end of §8.3. To use condition (1) as defining sufficiency, we observe that the likelihood function is the same for $X = 0$ and $X = 2$ (the rows that define it are proportional); it is different for $X = 1$ and $X = 3$. Thus, if you know the value $|X - 1|$, you know the likelihood function.

8-R4 (a) $P(X = 0) = f(0 \mid \theta_1)g(\theta_1) + f(0 \mid \theta_2)g(\theta_2) + \cdots$

$$= .2 \times .2 + .2 \times .4 + .4 \times .3 + .2 \times .1 = .26.$$

(b) The odds on values of θ are proportional to the terms in the above sum, or $4{:}8{:}12{:}2$. The probability of θ_3 is $12/(4+8+12+2) = 12/26$.

8-R5 (a) $E(\overline{X}/2) = (E\overline{X})/2 = \mu/2 = \theta$, and $\mathrm{var}(\overline{X}/2) = \dfrac{\mathrm{var}\,\overline{X}}{4} = \dfrac{\sigma^2}{4n} = \dfrac{\theta^2}{2n}$.

(b) The central limit theorem tells us that T is asymptotically normal; the parameters are those found in (a).

8-R6 (a) $L(\lambda) = \prod \lambda^2 e^{-\lambda x_i} = \lambda^{2n} e^{-\lambda\Sigma X_i}$ (proportional to the product of the marginal p.d.f.'s). When $n=10$ and $\sum X_i = 12$, $L(\lambda) = \lambda^{20} e^{-12\lambda}$, for $\lambda > 0$.

(b) $h(\lambda\,|\,\mathrm{data}) \propto g(\lambda)L(\lambda) \propto \lambda^{20} e^{-13\lambda}$, $\lambda > 0$. The proportionality constant is $13^{21}/20!$ [This is "$\lambda^\alpha/\mathrm{Gam}(\alpha)$", where "$\lambda$" $= 13$, "α" $= 21$.]

(c) The population distribution is $\mathrm{Gam}(2, \lambda)$, and by Problem 6-21, the sum of n independent observations is $\mathrm{Gam}(2n, \lambda)$.

8-R7 For $X \sim \mathcal{U}(0, 1)$, $EX = 1/2$, $\mathrm{var}\,X = 1/12$. So $E\overline{X} = 1/2$, and $\mathrm{var}\,\overline{X} = \dfrac{\mathrm{var}\,X}{n} = \dfrac{1}{60^2}$. Then, by the central limit theorem, $\overline{X}$ is approximately normal with these parameters, and
$$P(\overline{X} > .54) = 1 - \Phi\Big(\frac{.54 - .5}{1/60}\Big) = \Phi(-2.4).$$

8-R8 $np = 22.5$, $npq = 20.25$, and $n\hat{p} \approx \mathcal{N}(22.5, 20.25)$, or $\hat{p} \approx \mathcal{N}(.10, .02^2)$:
$$P(\hat{p} < .05) \doteq \Phi\Big(\frac{.05 - .10}{.02}\Big) = \Phi(-2.5).$$

8-R9 (a) $L(\theta) = \theta^{-n} e^{-\Sigma X_i^2/(2\theta)} = g(\theta, \sum X_i^2)$, so $\sum X_i^2$ is sufficient for θ.

(b) $E(X^2) = \mathrm{var}\,X + \mu^2 = 2\theta(1 - \pi/4) + \pi\theta/2 = 2\theta$. So
$$ET = E(\sum X_i^2) = nE(X^2) = 2n\theta.$$

(c) If $Y = X^2/\theta$, $f_Y(y) = f_X(\sqrt{\theta y})\cdot\dfrac{\sqrt{\theta}}{2\sqrt{y}} = \dfrac{\sqrt{\theta y}}{\theta}e^{-y/2}\cdot\dfrac{\sqrt{\theta}}{2\sqrt{y}} = \tfrac{1}{2}e^{-y/2}$

for $y > 0$. And this is the p.d.f. of $\mathrm{chi}^2(2)$, so $\sum X_i^2/\theta \sim \mathrm{chi}^2(2n)$.

8-R10 For large samples, $\bar{X}_1 - \bar{X}_2 \approx N(\mu_1 - \mu_2, .25^2)$, where

$$.25^2 = \frac{9}{400} + \frac{4}{100} = \frac{\sigma_1^2}{n_1} + \frac{\sigma_2^2}{n_2} = \text{var}(\bar{X}_1 - \bar{X}_2)$$

Then,

(a) If $\mu_1 = \mu_2$, $\bar{X}_1 - \bar{X}_2 \approx N(0, .0625)$, and

$$P(|\bar{X}_1 - \bar{X}_2| > .5 \mid \mu_1 - \mu_2 = 0) \doteq 2\Phi\left(\frac{-.5}{.25}\right) = 2\Phi(-2) = .0456.$$

(b) If $\mu_1 - \mu_2 = .2$, $\bar{X}_1 - \bar{X}_2 \approx N(.2, .0625)$, and

$$P(|\bar{X}_1 - \bar{X}_2| > .5 \mid \mu_1 - \mu_2 = .2)$$

$$\doteq 1 - \left\{\Phi\left(\frac{.5 - .2}{.25}\right) - \Phi\left(\frac{-.5 - .2}{.25}\right)\right\} = \Phi(-2.8) + \Phi(-1.2) = .1177.$$

8-R11 (a) The likelihood is proportional to the joint p.d.f. of the Y_i's, which are distributed as $N(\beta x_i, \sigma^2)$:

$$\prod (\sigma^2)^{-1/2}\exp\left\{-\frac{1}{2\sigma^2}(Y_i - \beta x_i)^2\right\} = (\sigma^2)^{-n/2}\exp\left\{-\frac{1}{2\sigma^2}\sum(Y_i - \beta x_i)^2\right\}$$

(b) The responses Y_i appear only in the exponent of the likelihood:

$$\sum_i (Y_i - \beta x_i)^2 = \sum_i Y_i^2 - 2\beta \sum x_i Y_i + \beta^2 \sum x_i^2.$$

The likelihood depends on the data via the statistics $\sum Y_i^2$ and $\sum x_i Y_i$, so these are sufficient statistics.

8-R12 (a) $F(x) = x$ and $f(x) = 1$, $0 < x < 1$; with $Y = X_{(k)}$, (1) of §8.6 is

$$f_Y(y) = \binom{n}{k-1,\ n-k} y^{k-1}(1-y)^{n-k},\ 0 < y < 1.$$

To find the mean, we integrate (as usual) $y f_Y(y)$:

$$EY = \binom{n}{k-1,\ n-k} \int_0^1 y^k(1-y)^{n-k}\,dy.$$

The integral is a beta function, evaluated using gamma functions

$$EY = \frac{n!}{(k-1)!(n-k)!}\,B(k+1,\ n-k+1).$$

$$= \frac{n!}{(k-1)!(n-k)!} \times \frac{k!(n-k)!}{(n+1)!} = \frac{k}{n+1}.$$

(b) $E[X_{(k+1)} - X_{(k)}] = \frac{k+1}{n+1} - \frac{k}{n+1} = \frac{1}{n+1}.$

8-R13 (a) $L(\lambda) = \prod f(X_i \mid \lambda) = \lambda^n e^{-\lambda \Sigma X_i}, \lambda > 0.$

(b) $h(\lambda \mid \text{data}) \propto g(\lambda)L(\lambda) \propto \lambda^{n+\alpha-1}e^{-(\Sigma X_i + \beta)\lambda}$. To make this into the posterior density (complete with constant factor), we divide it by its integral over $\lambda > 0$. Then, to get the unconditional p.d.f. of Y, we find the mean of $f(y \mid \lambda)$ with respect to the posterior density:

(c) $f^*(y) = E_h[f(y \mid \lambda)] = \dfrac{\displaystyle\int_0^\infty \lambda e^{-\lambda y}\lambda^{n+\alpha-1}e^{-(\Sigma X_i + \beta)\lambda}\,d\lambda}{\displaystyle\int_0^\infty \lambda^{n+\alpha-1}e^{-(\Sigma X_i + \beta)\lambda}\,d\lambda}$

$= \dfrac{\dfrac{\Gamma(n+\alpha+1)}{(y+\Sigma X_i + \beta)^{n+\alpha+1}}}{\dfrac{\Gamma(n+\alpha)}{(\Sigma X_i + \beta)^{n+\alpha}}} = \dfrac{n+\alpha}{\Sigma X_i + \beta} \times \left\{ \dfrac{1}{1+\dfrac{y}{\Sigma X_i + \beta}} \right\}^{n+\alpha+1}$

(d) Using the predictive p.d.f. from (c), we let $c = \beta + \sum X_i$, and find the expected value as the integral of $yf^*(y)$:

$$E_{f^*}(Y) = \frac{n+\alpha}{c}\int_0^\infty \frac{y\,dy}{(1+y/c)^{n+\alpha+1}}.$$

Make the substitution $u = 1 + y/c$, $y = c(u-1)$, $dy = c\,du$:

$$\frac{n+\alpha}{c}\int_1^\infty \frac{c^2(u-1)\,du}{u^{n+\alpha+1}} = c(n+\alpha)\left\{\int_1^\infty u^{-(n+\alpha)}du - \int_1^\infty u^{-(n+\alpha+1)}du\right\}$$

$$= \frac{c}{n+\alpha-1} = \frac{\beta+\sum X_i}{n+\alpha-1} = \frac{\beta/n + \overline{X}}{1+(\alpha-1)/n}.$$

As n becomes infinite, the strong law of large numbers says that $\overline{X}$ tends to the population mean with probability 1, as does this fraction.

8-R14 (a) Let $Y = X_{(k)}$, and $V = F(Y)$, $Y = F^{-1}(V)$, and $\dfrac{dy}{dv} = 1 / \dfrac{dv}{dy} = \dfrac{1}{f(y)}$.

Then,

$$f_V(v) = f_Y[F^{-1}(v)]\cdot\frac{dy}{dv} = f_Y[F^{-1}(v)]\cdot\frac{1}{f[F^{-1}(v)]}.$$

Again, from (1) of §8.6 for f_Y, we find that $f_V(v)$ is given by

$$\binom{n}{k-1,\,n-k}\{F[F^{-1}(v)]\}^{k-1}\{1-F[F^{-1}(v)]\}^{n-k}f[F^{-1}(v)]\cdot\frac{1}{f[F^{-1}(v)]}$$

$$=\binom{n}{k-1,\,n-k}v^{k-1}(1-v)^{n-k}\cdot 1,\ 0 < v < 1.$$

(b) From Problem 8-R12, we have $E(V_k) - E(V_{k-1}) = \frac{1}{n+1}.$

8-R15 (a) $\displaystyle\int_0^1 \frac{dp}{\sqrt{p(1-p)}} = \int_0^1 p^{-1/2}(1-p)^{-1/2}dp = B(\tfrac{1}{2},\tfrac{1}{2}) = \frac{\Gamma(\tfrac{1}{2})\Gamma(\tfrac{1}{2})}{\Gamma(\tfrac{1}{2}+\tfrac{1}{2})} = \pi.$

We then divide the integrand by the constant π to make it a p.d.f.

(b) $P(S\,|\,p) = p$, and

$$P(S) = E_g[P(S\,|\,p)] = \int_0^1 P(S\,|\,p)g(p)\,dp = \tfrac{1}{\pi}\int_0^1 p\cdot p^{-1/2}(1-p)^{-1/2}dp$$

The last integral is a beta function, so

$$P(S) = \tfrac{1}{\pi}\cdot B(3/2, 1/2) = \tfrac{1}{\pi}\cdot\frac{\tfrac{1}{2}\Gamma(\tfrac{1}{2})\Gamma(\tfrac{1}{2})}{\Gamma(2)} = \tfrac{1}{2}.$$

(c) After a success, $L(p) = p$; and $h(p\,|\,S) \propto p\cdot g(p) = \dfrac{p^{1/2}}{(1-p)^{1/2}}.$ Then the probability of success at a new trial is

$$P(S) = E_h[P(S\,|\,p)] = \frac{\displaystyle\int_0^1 p\,\frac{p^{1/2}}{(1-p)^{1/2}}dp}{\displaystyle\int_0^1 \frac{p^{1/2}}{(1-p)^{1/2}}dp} = \frac{B(5/2, 1/2)}{B(3/2, 1/2)} = \frac{\tfrac{3}{2}\cdot\tfrac{1}{2}\pi\Gamma(2)}{\tfrac{1}{2}\pi\Gamma(3)} = \frac{3}{4}.$$

[That is, the predictive distribution for the new trial is Ber(3/4).]

CHAPTER 9

9-R1 $\frac{18}{\sqrt{n}} \leq 1.2$, $\sqrt{n} \geq \frac{18}{1.2} = 15$, so $n = 225$ will do the trick.

9-R2 (a) With p unknown, we prepare for the worst case, $p = .5$:
$$\sqrt{\frac{.5 \times .5}{n}} = \sqrt{.25/n} \leq .015, \text{ or } .25/n \leq .015^2, \text{ or } n \geq 1{,}112.$$

(b) $\sqrt{.3 \times .7/1112} = .01374$; the limits are $.3 \pm 1.96 \times .01374$ or $.3 \pm .0269$.

9-R3 $\sqrt{\dfrac{9^2}{100} + \dfrac{12^2}{100}} = \sqrt{\dfrac{15^2}{100}} = 1.5.$

9-R4 $L(\theta) = \theta X^{\theta - 1}$, so $\log L = \log \theta + (\theta - 1) \log X$, $\dfrac{\partial \log L}{\partial \theta} = \dfrac{1}{\theta} + \log X$.
We need $\mathrm{var}(\log X)$; let $Y = -\log X$. Then
$$f_Y(y) = f_X(e^{-y})e^{-y} = \theta e^{-\theta y}, \ y > 0.$$
So, $-\log X$ is $\mathrm{Exp}(\theta)$, and $\mathrm{var}(\log X) = 1/\theta^2 = I(\theta)$.

9-R5 (a) $E(X^2) = \mathrm{var}\, X = \theta$, so $ET = \frac{1}{n}\sum X_i^2 = \theta$, and T is unbiased—and also sufficient, and therefore efficient.

(b) $\dfrac{X_i - 0}{\sqrt{\theta}} \sim \mathcal{N}(0, 1)$, so $\sum\limits_1^n X_i^2/\theta \sim \mathrm{chi}^2(n)$. This is pivotal because the distribution does not involve the parameter θ.

9-R6 (a) $L = e^{-\lambda}\lambda^X$, $\log L = -\lambda + X \log \lambda$, $\dfrac{\partial^2}{\partial \lambda^2}\log L = -\dfrac{X}{\lambda^2}$. The information $I(\lambda)$ in one observation is then $E(X/\lambda^2) = 1/\lambda$.

(b) $n \times I(\lambda) = n/\lambda$.

(c) $\mathrm{var}\,\overline{X} = \frac{1}{n}\mathrm{var}\, X = \frac{\lambda}{n} = \dfrac{1}{I_n(\lambda)}$, so $\overline{X}$ is efficient; its efficiency is 1.

(d) $\overline{X}$ is always unbiased in estimating the mean (λ), and its variance tends to 0; so it is consistent in mean square, hence consistent.

9-R7 (a) $L(\lambda) = e^{-n\lambda}\lambda^{\sum X_i} = e^{-10\lambda}\lambda^9$, and

$$h(\lambda \mid \text{data}) \propto g(\lambda)L(\lambda) \propto e^{-\lambda} \times e^{-10\lambda}\lambda^9 = e^{-11\lambda}\lambda^9, \ \lambda > 0.$$

This is the p.d.f. of Gam$(10, 11)$, whose posterior mean is "α/λ" $= 10/11$.

(b) With absolute error loss, the Bayes estimate is the posterior median.

We look for y such that $\sum_0^9 \dfrac{(11y)^k}{k!} e^{-11y} = .5$. Table IV shows that this

sum is $.5$ at $c \doteq 9.65 = 11y$, so $y = 9.65/11 = .877$.

9-R8 (a) T is unbiased (8-R5): m.s.e. $= E[(T - 1/\lambda)^2] = \text{var } T = \dfrac{1}{2n\lambda^2}$ (8-R5).

(b) $L(\lambda) = \lambda^{2n}e^{-\lambda\sum X_i}, \ \lambda > 0$; $\log L = 2n\log\lambda - \lambda\sum X_i$, and

$\dfrac{\partial}{\partial\lambda}\log L = \dfrac{2n}{\lambda} - \sum X_i$, which vanishes at $\hat{\lambda} = \dfrac{2}{\bar{X}}$, the m.l.e. of λ. The

m.l.e. of $\theta = 1/\lambda$ is $\hat{\theta} = 1/\hat{\lambda} = \bar{X}$.

(c) Write EX in terms of the parameter: $\mu = 1/\lambda$; solve for λ: $\lambda = 1/\mu$,
and replace μ by the corresponding sample moment: $1/\bar{X}$. This is the
"method of moments estimator."

(d) $L(\lambda) = \lambda^{20}e^{-12\lambda}$, and $g(\lambda) = \lambda e^{-\lambda}$, so $h(\lambda \mid \text{data}) \propto \lambda^{21}e^{-13\lambda}, \ \lambda > 0.$

This p.d.f. is Gam$(22, 13)$, with mean $22/13 = 1.692$. $(\hat{\lambda} = \dfrac{2}{\bar{X}} = 1.667)$

9-R9 (a) $L(\theta) = \dfrac{1}{\theta^n}\exp\left\{-\dfrac{1}{2\theta}\sum X_i^2\right\}, \ \theta > 0$; $\log L = -n\log\theta - \dfrac{1}{2\theta}\sum X_i^2$, with

derivative $-\dfrac{n}{\theta} + \dfrac{1}{2\theta^2}\sum X_i^2$, which vanishes at $\hat{\theta} = \dfrac{1}{2n}\sum X_i^2$, the m.l.e.

(b) Solve $\mu = \sqrt{\pi\theta/2}$ for θ in terms of μ: $\theta = 2\mu^2/\pi$, and replace μ by
the corresponding sample moment, $\bar{X}$.

(c) Because $X = \sqrt{\theta(Z_1^2 + Z_2^2)}$, it follows that $X^2/\theta = Z_1^2 + Z_2^2 \sim \text{chi}^2(2)$.

(d) The m.l.e. is unbiased: $E\hat{\theta} = \dfrac{1}{2n}\sum E(X_i^2) = \dfrac{n}{2n}E(X^2) = \theta$ [Problem

8-R9(b)]. The m.s.e. is therefore just the variance:

$$\text{var } \hat{\theta} = \left(\dfrac{1}{2n}\right)^2 \sum \text{var } X_i^2 = \dfrac{1}{4n}\text{var } X^2.$$

From (c) we know that $\text{var}(X^2/\theta) = 2 \times \text{d.f.} = 4$, so $\text{var}(X^2) = 4\theta^2$, and
m.s.e. $= \theta^2/n$.

9-R10 It's actually a bit easier to do (b) first:

(b) If $ET = \theta$, then $E(T^2) = \operatorname{var} T + (ET)^2 = \operatorname{var} T + \theta^2 > \theta^2$. So except for the trivial case where T is singular, T^2 is biased.

(a) If $\sqrt{T}$ were *not* biased in estimating $\sqrt{\theta}$, then [from (b)], it would follow that T would be biased as an estimate of θ, contrary to the assumption.

9-R11 Given that $\overline{D} \sim \mathcal{N}(\theta, 1.252)$, the 95% limits are $2.41 \pm 1.96 \times \sqrt{1.252}$, or $.217 < \mu < 4.603$.

(a) It is given that $\pi_{\mathrm{pr}} = 1/4$, and $\pi_{\mathrm{data}} = 1/1.252$. This means that

$$\pi_{\mathrm{po}} = \frac{1}{4} + \frac{1}{1.252} = .25 + .7987 = 1.0487 = \frac{1}{.9765^2}.$$

The posterior mean is a weighted average of the sample and prior means, with weights proportional to the precisions of prior and data:

$$\mu_{\mathrm{po}} = \frac{.25}{1.0487} \times 4 + \frac{.7987}{1.0487} \times 2.41 = 2.789.$$

So, the posterior is $\mathcal{N}(2.789, .9765^2)$, and

$$P(.217 < \mu < 4.603) = \Phi\left(\frac{4.603 - 2.789}{.9765}\right) - \Phi\left(\frac{.217 - 2.789}{.9765}\right) = .963.$$

(b) Repeat the process in (a), but with 1/4 replaced by 1/16:

$$\pi_{\mathrm{po}} = \frac{1}{16} + \frac{1}{1.252} = .86122 = \frac{1}{1.07756^2}.$$

The posterior mean is a weighted average of the sample and prior means:

$$\mu_{\mathrm{po}} = \frac{1/16}{.86122} \times 4 + \frac{.7987}{.86122} \times 2.41 = 2.5254.$$

So, the posterior is $\mathcal{N}(2.5254, 1.07756^2)$, and

$$P(.217 < \mu < 4.603) = \Phi\left(\frac{4.603 - 2.5254}{1.07756}\right) - \Phi\left(\frac{.217 - 2.5254}{1.07756}\right) = .957.$$

CHAPTER 10

10-1R (a) To define a *P*-value, we find the area beyond $\overline{X} = .75$ in the distribution of $\overline{X}$ under the hypothesis $\mu = 0$ (evidence against any $\mu < 0$ is even stronger):

$$P(\overline{X} > .75 \mid \mu = 0) = 1 - \Phi\left(\frac{.75 - 0}{2}\right) = .354.$$

(b) Given $\nu = .5$, $\tau^2 = 1$, $\pi_{\text{data}} = 1/4$, $\pi_{\text{pr}} = 1$, we get $\pi_{\text{po}} = 1 + 1/4$ and $\sigma_{\text{po}}^2 = 1/\pi_{\text{po}} = 4/5 = .8$. Then

$$E_h(\mu) = \frac{1 \times .5 + \frac{1}{4} \times .75}{1 + \frac{1}{4}} = .55,$$

whence

$$P(\mu \leq 0 \mid \text{data}) = \Phi\left(\frac{0 - .55}{\sqrt{.8}}\right) = \Phi(-.615) = .269.$$

10-R2 (a) $P(X > 2 \mid H_0) = 1 - \Phi\left(\frac{2 - 0}{1}\right) = .0228.$

(b) $P(Y > 4 \mid H_0) = 1 - F_{\chi^2(1)}(4) = .045.$

10-R3 There are 3 patterns with R_- of 2 or smaller:

$$+ + + + + +$$
$$- + + + + +$$
$$+ - + + + +$$

so $P = 3/64$, where $64 = 2^6$ is the number of possible patterns.

10-R4 The data are: 1.2, 2.4, 1.3, 1.3, 0.0, 1.0, 1.8, 0.8, 4.6, 1.3.

(a) Without the 4.6, $\overline{X} = 1.233$, $S = .6576$

$$T = \frac{1.233 - 0}{.6576/\sqrt{9}} \doteq 5.63, \text{ 8 d.f., } P = .000.$$

Including the 4.6 gave us $T = 4.03$ (9 d.f.) and $P = .002$. What makes T larger, without the observation farthest from 0, is that the standard error is smaller; so 1.233 is larger in comparison. [The outlier 4.6 suggests that the population is skewed, and the *t*-test perhaps shouldn't be used.]

(b) Without the 4.6, observations are all positive; the probability of this (under H_0) is now $1/2^9 \doteq .002$—the *P*-value for the sign test. The signed rank test also gives this *P*-value, because there is only one sign sequence (out of 2^9) with $R_- = 0$.

10-R5 $\bar{X} = 30.7875$, $S = 6.530$, $T = \dfrac{30.7875 - 29.0}{6.53/\sqrt{8}} = .774$ (7 d.f.), $P > .2$:

The observed difference is less than the difference one typically observes when H_0 is true.

10-R7 $P(t) = 1 - F_T(t)$, and we know that $F_X(X) \sim \mathcal{U}(0, 1)$ for any random variable X. Thus, $F_T(T) \sim \mathcal{U}(0, 1)$, and (by reflection around .5) so is $1 - F_T(T)$.

10-R8   We know $(n-1)S^2/\sigma^2 \sim \text{chi}^2(n-1)$, so we write the P-value in terms of $8S^2/9$:

$$P(S > 4.206 \mid H_0) = P\left(\frac{8S^2}{9} > \frac{8 \times 4.206^2}{9} \mid \sigma^2 = 9\right).$$

This is the tail-area of $\text{chi}^2(8)$ beyond $8 \times 4.206^2/9 = 15.725$, about .046.

10-R9   (a) $\mu \sim \mathcal{N}(2.787, .9765^2)$: $P(\mu \leq 0) = \Phi\left(\dfrac{0 - 2.787}{.9765}\right) = \Phi(-2.854) = .002$.

(b) $\mu \sim \mathcal{N}(2.5254, 1.0776^2)$, and

$$P(\mu \leq 0) = \Phi\left(\frac{0 - 2.5254}{1.0776}\right) = \Phi(-2.344) = .0095.$$

10-R10 $L(p) = p^6(1-p)^4$, and $g(p) \propto p^2(1-p)^4$, so $h(p \mid \text{data}) \propto p^8(1-p)^8$. This p.d.f. is symmetric about $p = 1/2$, so $P_h(p \leq .5) = 1/2$.

10-R11 $\dfrac{L_0}{L_A} = \dfrac{e^{-\Sigma X_i}}{2^8 e^{-2\Sigma X_i}} = e^{\Sigma X_i}/2^8$.

(a) $L_0 = L_A$ if $e^{\Sigma X} = 256$, or $\sum X_i = \log 256 = 5.545$, or $\bar{X} = .693$.

(b) $L_A = 4L_0$ if $2^8 e^{-2\Sigma X_i} = 4e^{-\Sigma X_i}$, or $256 = 4e^{\Sigma X_i}$, or $\sum X_i = \log 64$.

CHAPTER 11

11-R1 (a) $\dfrac{L_0}{L_1} = \dfrac{e^{-\frac{1}{2}\Sigma X_i^2}}{e^{-\frac{1}{8}\Sigma X_i^2}} = e^{-\frac{3}{8}\Sigma X_i^2}$. This is small if $\sum X_i^2$ is large.

(b) $\sum_1^4 X_i^2/\sigma^2 \sim \mathrm{chi}^2(4)$—*not* 3 d.f., since μ is given, not estimated;

so, $\alpha = P\left(\sum_1^4 X_i^2 > K \mid \sigma^2 = 1\right) = P\left(\sum_1^4 X_i^2/1 > K\right) = .10$ if $K = 7.78$.

(c) $\beta = P\left(\sum_1^4 X_i^2 < K \mid \sigma^2 = 4\right) = P\left(\sum_1^4 X_i^2/4 < K/4\right) = F_{\chi^2(4)}(7.78/4)$.

Using the table of chi-square percentiles, we can only find this as being between .20 and .30.

(d) Let $T = \sum_1^n X_i^2$; then $T/\sigma^2 \sim \mathrm{chi}^2(n)$. We want n and K such that
$$\alpha = P(T > K \mid \sigma^2 = 1) = .01, \ \beta = P(T/4 < K/4 \mid \sigma^2 = 4) = .01.$$

With the tables we have, we can only approximate the solution: look in the table of chi-square percentiles for an n such that the 99th percentile is 4 times the first percentile. This is approximately the case for $n = 23$, where $K = 41.6$, and $K/4 = 10.4$.

11-R2 $f(x \mid \lambda) = e^{-\lambda}\lambda^x/x!$, so $L(\lambda) = e^{-n\lambda}\lambda^{\Sigma X_i}$, $\lambda > 0$.

(a) $\dfrac{L(1)}{L(2)} = \dfrac{e^{-5} \times 1^{\Sigma X_i}}{e^{-10} \times 2^{\Sigma X_i}} = \dfrac{e^5}{2^{\Sigma X_i}}$. This is small when $\sum X_i$ is large.

(b) $\sum_1^5 X_i \sim \mathrm{Poi}(5\lambda)$. When $\lambda = 1$, this is $\mathrm{Poi}(5)$, so we use Table IV with $m = 5$ to find
$$\alpha = P\left(\sum X_i > 7 \mid \lambda = 1\right) = 1 - P\left(\sum X_i \le 7 \mid m = 5\right) = 1 - .867.$$

When $\lambda = 2$, $\sum X_i \sim \mathrm{Poi}(10)$, so $\beta = P\left(\sum X_i \le 7 \mid \lambda = 2\right) = .220$.

11-R3 Under H_0, $\mu_{\bar{X}} = 5$, $\sigma_{\bar{X}}^2 = 1/4$. The power function is thus
$$\pi(\mu) = P(\mathrm{rej.}\ H_0 \mid \mu) = P(\bar{X} > 6 \mid \mu) = 1 - \Phi\left(\frac{6 - \mu}{1/2}\right) = \Phi(2\mu - 12).$$

11-R4 (a) $\alpha = P(S^2 > 4 \mid \sigma^2 = 2) = P(9S^2/2 > 9 \times 4/2) = .035$ [because $9S^2/2$

(that is, $(n-1)S^2/\sigma^2$) is chi$^2(9)$ when $\sigma^2 = 2$], from Table V.

(b) $\pi(\sigma^2) = P(S^2 > 4 \mid \sigma^2) = P(9S^2/\sigma^2 > 9 \times 4/\sigma^2) = 1 - F_{\chi^2(9)}(36/\sigma^2)$.

To draw the graph, use the table of chi-square percentiles (9 d.f.):

$\pi(\sigma^2)$	$36/\sigma^2$	σ^2
.99	2.09	17.22
.95	3.33	10.81
.90	4.17	8.63
.80	5.38	6.69
.50	8.34	4.32
.20	12.2	2.95
.10	14.7	2.45
.05	16.9	2.13
.01	21.7	1.66

Now plot power vs. σ^2.

11-R5 The table below repeats the three distributions of Z, with corresponding prior probabilities alongside the θ's. With these, we find $f(z)$:

		z_1	z_2	z_3	z_4
.2	θ_1	.2	.3	.1	.4
.5	θ_2	.6	.1	.1	.2
.3	θ_3	.3	0	.4	.3

$f(z_1) = .2 \times .2 + .5 \times .6 + .3 \times .3 = .43$

$f(z_2) = .2 \times .3 + .5 \times .1 + .3 \times 0 = .11$

$f(z_3) = .2 \times .1 + .5 \times .1 + .3 \times .4 = .19$

$f(z_4) = .2 \times .4 + .5 \times .2 + .3 \times .3 = .27$

(a) Given $Z = z$, the posterior probabilities of the θ's are proportional to the terms in the corresponding sum. Thus, given $Z = z_1$, they are proportional to 4:30:9, or 4/43, 30/43, 9/43. Similarly, if $Z = z_2$: 6/11, 5/11, 0; if $Z = z_3$: 2/19, 5/19, 12/19; and if $Z = z_4$, they are 8/27, 10/27, 9/27.

(b) To find the Bayes action, we look at the expected posterior losses for each possible action and choose the one with the smaller mean loss. So, if we were to reject H_0, the losses are 4, 0, 0 for θ_1, θ_2, θ_3, and the expected posterior losses (which depend on z) would be: if $Z = z_1$: $4 \times P_h(\theta_1) = 16/43$; if $Z = z_2$: $4 \times P_h(\theta_1) = 24/11$;

if $Z = z_3$: $4 \times P_h(\theta_1) = 8/19$; and if $Z = z_4$: $4 \times P_h(\theta_1) = 32/27$.

If we accept H_0, the losses are 0, 2, 1 for θ_1, θ_2, θ_3, and mean losses are: if $Z = z_1$, $P_h(\theta_1) + 2 \times P_h(\theta_2) = 69/43$; (similarly) if $Z = z_2$, $10/11$; if $Z = z_3$, $22/19$; and if $Z = z_4$, $29/27$.

So, we reject H_0 if $Z = z_1$ $[16/43 < 69/43]$, or $Z = z_3$ $[8/19 < 22/19]$, and otherwise accept H_0 [because $24/11 > 10/11$, and $32/27 > 29/37$]—to minimize expected posterior losses.

11-R6 The likelihood function is $L(\theta) = \frac{1}{\theta^n} e^{-\Sigma X_i^2/(2\theta)}$, $\theta > 0$. The maximum occurs at $\widehat{\theta} = \frac{1}{2n}\Sigma X_i^2$, and $\sup\limits_\theta L(\theta) = L(\widehat{\theta}) = \frac{e^{-n}}{(\widehat{\theta})^n}$. Under the null hypothesis $(\theta = 1)$, the (maximum) likelihood is $L(1) = e^{-n\widehat{\theta}}$. And then

$$\Lambda = \frac{L(1)}{L(\widehat{\theta})} = e^n(\widehat{\theta}\, e^{-\widehat{\theta}})^n.$$ This is small if $\widehat{\theta}\, e^{-\widehat{\theta}}$ is small, which will occur

when $\widehat{\theta}$ is either very small or very large. Figure 4 shows a plot of ue^{-u}, and shows critical values A and B for $\widehat{\theta}$, corresponding to $\widehat{\theta}\, e^{-\widehat{\theta}} < .1$.

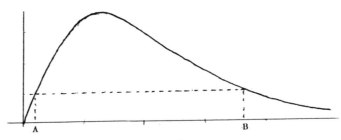

Figure 4

11-R7 $L(\beta, \sigma^2) = (\sigma^2)^{-n/2}\exp\left\{-\frac{1}{2\sigma^2}\Sigma(Y_i - \beta x_i)^2\right\}$ (from Problem 8-R11).

$\frac{\partial}{\partial\beta}\log L = -\frac{1}{\sigma^2}\Sigma(-x_i)(Y_i - \beta x_i)$, which vanishes at $\widehat{\beta} = \frac{\Sigma x_i Y_i}{\Sigma x_i^2}$, and

$\frac{\partial}{\partial\sigma^2}\log L = -\frac{n}{2\sigma^2} + \frac{1}{2\sigma^4}\Sigma(Y_i - \beta x_i)^2$. Both derivatives vanish at

$(\widehat{\theta}, \widehat{\sigma}^2)$, where $\widehat{\sigma}^2 = \frac{1}{n}\Sigma(Y_i - \widehat{\beta}x_i)^2$, and this is the maximum point. The maximum value is $L(\widehat{\beta}, \widehat{\sigma}^2) = (e\widehat{\sigma}^2)^{-n/2}$.

We also need the maximum of L when $\beta = 0$, or of

$$L(0, \sigma^2) = (\sigma^2)^{-n/2}\exp\left\{-\frac{1}{2\sigma^2}\Sigma Y_i^2\right\}.$$

This occurs at $\hat{\sigma}_0^2 = \sum Y_i^2/n$, so

$$\Lambda = \frac{L(0, \hat{\sigma}_0^2)}{L(\hat{\beta}, \hat{\sigma}^2)} = \left\{\frac{\hat{\sigma}_0^2}{\hat{\sigma}^2}\right\}^{-n/2}.$$

11-R8 (a) $L(p_1, p_2, p_3) = p_1^{15} p_2^{15} p_3^{30}$, so $L(\frac{1}{3}, \frac{1}{3}, \frac{1}{3}) = (\frac{1}{3})^{60}$, $L(\frac{1}{4}, \frac{1}{4}, \frac{1}{2}) = (\frac{1}{4})^{30}(\frac{1}{2})^{30}$.

(b) $\Lambda = \dfrac{(\frac{1}{3})^{60}}{(\frac{1}{4})^{30}(\frac{1}{2})^{30}} = \dfrac{2^{90}}{3^{60}}$, and $-2\log\Lambda = 7.067$ (2 d.f.), so $P = .029$.

CHAPTER 12

12-R1 Assuming $p_1 = p_2 = p$, we estimate p by taking the ratio of the total number of successes to the total number of trials: $\hat{p} = 85/416$. The s.d. of the difference in proportions is then estimated to be

$$\text{s.e.} = \sqrt{\frac{85}{416} \cdot \frac{331}{416}\left(\frac{1}{200} + \frac{1}{216}\right)} = .039567.$$

Then

$$Z = \frac{29/200 - 56/216}{.039567} = -2.887, \quad P \doteq .002.$$

12-R2 (a) The samples are independent, so we use the two-sample Wilcoxon rank-sum statistic:

$R_1 = 1 + 2 + 3 + 5 + 6 = 17$, $n_1 = n_2 = 5$, so $P = .016$ (Table VII).

(b) $R_1 \le 17$ for four sequences: 1 1 1 1 1 2 2 2 2 2
                                    1 1 1 1 2 1 2 2 2 2
                                    1 1 1 2 1 1 2 2 2 2
                                    1 1 1 1 2 2 1 2 2 2

Thus, $P = \dfrac{4}{\binom{10}{5}} = \dfrac{1}{63} \doteq .01587.$

12-R3 (a) The difference of independent normal variables is normal, and $\bar{X}_1$ and $\bar{X}_2$ are each (being sample from normal populations) normal. Thus, the difference $\hat{\delta}$ is normal, with mean $E(\bar{X}_1 - \bar{X}_2) = \mu_1 - \mu_2$ and

$$\text{var}(\bar{X}_1 - \bar{X}_2) = \text{var}(\bar{X}_1) + \text{var}(\bar{X}_2) = \sigma^2/n_1 + \sigma^2/n_2.$$

(b) var $\hat{\delta} = 1/8 + 1/8 = (1/2)^2$, so

$$\pi(\delta) = P(|\hat{\delta}| > 1) = 1 - [P(-1 < \hat{\delta} < 1)] = 1 - \Phi\left(\frac{1-\delta}{1/2}\right) + \Phi\left(\frac{-1-\delta}{1/2}\right).$$

and $\pi(2) = 1 - \Phi(-2) + \Phi(-6) = 1 - .0228 - .0000 = .9772.$

12-R4 If $p_1 = p_2 = p$, the m.l.e. of p is $\hat{p} = 980/2000 = .49$. We test using the

approximate Z-score: $Z = \dfrac{.50 - .48}{\sqrt{.49 \times .51 \times (.001 + .001)}} = .8946; \quad P \doteq .185.$

12-R5 In numerical order: 31, 38, **49**, 64, **73**, 74, 74, **75**, **76**, 90, 100, **115**, with the observations of the smaller sample in boldface. These have ranks 3, 5, 8, 9, 12, with sum $R_1 = 37$. In Table VII ($m = 5$, $n = 7$), we find

$$P = P(R_1 \geq 37) = P(65 - R_1 \leq 28) > .216.$$

The two-sample t-statistic is $T = \dfrac{77.6 - 67.286}{24.72\sqrt{1/5 + 1/7}} = .71$ (10 d.f.).

12-R6 Ranks are R_S: 4.5, 4.5, 10, 10, 13, 16, 16, 22, 25, 26, 27, 28
$ R_N$: 1, 2, 4.5, 4.5, 7, 8, 10, 13, 13, 16, 18, 19, 20, 21, 23, 24

Taking these as sample of sizes 12 and 16, we find means $\overline{X}_S = 16.833$,

$S_S = 8.6296$, $\overline{X}_N = 12.75$, $S_N = 7.670$, $S_p^2 = \dfrac{11 \times 8.63^2 + 15 \times 7.67^2}{26}$

or $S_p^2 = 8.09^2$. The two-sample t-statistic is then

$$T = \frac{16.833 - 12.75}{8.09\sqrt{(1/12 + 1/16)}} = 1.322 \quad (26 \text{ d.f.}), \quad P \doteq .1.$$

12-R7 The data are paired, so we use *one*-sample tests on the *differences*, D_i:

$$.43, \ -.02, \ .36, \ .13, \ .19, \ .16, \ .15, \ -.01, \ .23, \ -.11.$$

(a) The mean is $\overline{D} = .151$, and the s.d. is $S_B = .1677$. With these,

$$T = \frac{.151 - 0}{.1677/\sqrt{10}} = 2.846 \quad (9 \text{ d.f.}), \text{ and } P = .0095.$$

(b) The differences with smallest magnitudes are $-.01$, $-.02$, $-.11$, which happen to be all of the negative ones. Thus, $R_- = 1 + 2 + 3 = 6$. Table VI shows P to be .014.

(c) The are 3 negative differences among the 10, so the P-value is the probability of 3 or fewer negatives when the probabilities of $+$ and $-$ are (each) $p = .5$: $P[\#(-\text{'s}) \leq 3 \mid p = .5] = .1719$ (Table I).

12-R8 We use a large sample Z-test for a zero mean difference:

$$Z = \frac{32 - 0}{\sqrt{\dfrac{87^2}{2046} + \dfrac{77^2}{1668}}} = 11.8.$$

The P-value is extremely small—the observed difference is "highly significant" by most anyone's standard.

12-R9 As in 12-R4, a large sample Z-test for a zero difference in population proportions. The percentages given have been rounded off; for gays, 30 percent undoubtedly came from $20/66 = .303$. For others, 14 percent could have been rounded from 25 or 26 out of 182; we'll say it's $26/182$. The estimate of the common p (under H_0) is $46/248$, and

$$Z = \frac{20/66 - 26/182}{\sqrt{\frac{46}{248} \cdot \frac{200}{248}\left(\frac{1}{182} + \frac{1}{66}\right)}} = \frac{.159}{.0561} = 2.834, \; P = .0023.$$

12-R10 (a) Since the prior is uniform on the unit square, the probability that $p_1 > p_2$ (which defines a triangular region, half of the square) is $1/2$.

(b) Given that a single observation on population 1 is a success, and a single observation on population 2 is a failure, the likelihood function is

$$L(p_1, p_2) = p_1(1 - p_2), \; 0 < p_i < 1 \; (i = 1, 2).$$

Since the prior is constant, the posterior joint p.d.f. is proportional to the likelihood: $h(p_1, p_2 \mid \text{data}) \propto p_1(1 - p_2)$. With this, calculate the desired probability by integrating the posterior p.d.f. over the triangle where $p_1 > p_2$:

$$P_h(p_1 > p_2) = \frac{\int_0^1 \left\{ \int_0^{p_1} p_1(1 - p_2)\, dp_2 \right\} dp_1}{\int_0^1 \left\{ \int_0^1 p_1(1 - p_2)\, dp_2 \right\} dp_1}.$$

[The reciprocal of the denominator is the proportionality constant for h.] In the denominator, the p_1 factors out of the inner integral, and then the p_2 integral factors out of the p_1 integral, so that the denominator is the product of two single integrals. In the numerator, integrate first on p_2:

$$P_h(p_1 > p_2) = \frac{\int_0^1 p_1\left(p_1 - \tfrac{1}{2}p_1^2\right) dp_1}{\int_0^1 p_1\, dp_1 \int_0^1 (1 - p_2)\, dp_2} = \frac{5/24}{1/4} = \frac{5}{6}.$$

12-R11 The data are *paired*, so we deal with this by applying a one-sample test to the sample of differences:

$$-165, 300, 550, -39, -124, 33, -155, -228, 86, 123, 70, 262, 243, 46.$$

We first find $\bar{D} = 71.57$, and $S_D = 214.7$. Then $T = \dfrac{71.57}{214.7/\sqrt{14}} \doteq 1.25$, and $P = .117$—not much of an argument against H_0. (We could also have found the signed-rank statistic:

$$R_- = 9 + 2 + 7 + 8 + 10 = 36, \quad P > .134.)$$

12-R12 The sample statistics are as follows:

	n	*Mean*	*S.D.*
Nonsmokers	38	7.218	1.3460
Smokers	28	6.807	1.0143

We could use the two-sample T-statistic (64 d.f.), but perhaps the d.f. is large enough that a Z-statistic is adequate:

$$Z = \frac{7.218 - 6.807}{\sqrt{\dfrac{1.3460^2}{38} + \dfrac{1.0143^2}{28}}} \doteq 1.4.$$

With $S_p^2 = 1.4814$, we find that $T = 1.356$. Either way, the evidence is not overwhelming against H_0. (Moreover, we can say nothing as to whether, if the difference is "real," it is caused by the smoking—this was not a controlled experiment.)

CHAPTER 13

13-R1 If we assume equally likely sides, the expected frequencies in 72 tosses would be 18 for each side. Then

$$\chi^2 = \frac{6^2}{18} + \frac{2^2}{18} + \frac{2^2}{18} + \frac{6^2}{18} = \frac{40}{9} = 4.444.$$

With 4 categories, d.f. $= 4 - 1 = 3$, and $P \doteq .22$.

13-R2 A rough sketch shows the maximum deviation at .60, where $F_0 - F_n$ is 0.4. Without a sketch, we could begin to look at the table of values:

$x_{(i)}$	$F_0(x_{(i)})$	$F_n(x_{(i)})$	a_i	b_i
.24	.24	.10	.14	.24
.35	.35	.20	.15	.25
.60	.60	.30	.30	.40
.64	.64	.50	.14	.34
.70	.70	.60	.10	.20

And from this point on, the distances a and b get smaller; $D_n = .4$, and the P-value (Table IXa) is about .05.

13-R3 The data are paired, and we treat them as comprising a sample of size 40, cross-classified as follows:

	N	S	
N	24	10	34
S	3	3	6
	27	13	40

To test equality of the proportions of smokers among mothers and among fathers, we use McNemar's test, with the statistic based only on the "off-diagonal" frequencies, 3 and 10:

$$Z = \frac{(10 - 3)^2}{13} \doteq 3.8, \quad P = .0001 \text{ (Table 2a)}.$$

13-R4 (a) Since the sample sizes are equal, (estimated) expected frequencies are equal for a given response: 510 for Yes, 310 for No, 180 for Undecided. Then,

$$\chi^2 = \frac{10^2}{510} + \frac{10^2}{510} + \frac{10^2}{310} + \frac{10^2}{310} + \frac{20^2}{180} + \frac{20^2}{180} \doteq 5.48.$$

With d.f. $= (3 - 1)(2 - 1) = 2$, we find $P \doteq .065$ (Table Vb).

(b) See (6) in §13.6 for Λ, or the box that follows it, for $-2\log\Lambda$:

$$-2\log\Lambda = 4000\log 2000 + 2\times[500\log 500 + 520\log 520 + 300\log 300$$
$$320\log 320 + 200\log 200 + 160\log 160]$$
$$-2\times 2000\log 1000 - 2040\log 1020 - 1240\log 620 - 720\log 360$$
$$= 5.491 \text{ (again 2 d.f.)}.$$

13-R5 Given frequencies Y_i for categories with probabilities p_i, the likelihood function is $L(p_1, ..., p_4) = p_1^{Y_1} p_1^{Y_2} p_1^{Y_3} p_1^{Y_4}$. Here the cell probabilities depend on the parameter θ:

(a) $L(\theta) = [\theta^2]^{32}[\theta(1-\theta)]^{16}[\theta(1-\theta)]^{20}[(1-\theta)^2]^{32} = \theta^{100}(1-\theta)^{100}$.

(b) This is a binomial likelihood, maximized at $\hat{\theta} = 100/(100+100) = .5$.

(c) With the one parameter θ estimated in (b) to be $1/2$, the estimated expected cell frequencies in a sample of 100 are each 25. Then,
$$\chi^2 = \frac{7^2}{25} + \frac{9^2}{25} + \frac{5^2}{25} + \frac{7^2}{25} = \frac{204}{25} = 8.16, \text{ with d.f.} = 4 - 1 - 1 = 2, \text{ and}$$
$P \doteq .017$.

13-R6 The rankits for $n = 10$ (from Table XII) are:

$$-1.539, -1.001, -.656, -.376, -.123, .123, .376, .656, 1.001, 1.539.$$

Pair these with the ordered data: 0, .8, 1, 1.2, 1.3, 1.3, 1.3, 1.8, 2.4, 4.6 and find the correlation coefficient, $r = .8889$, and $W = r^2 = .7904$. Table XIV shows $P = .01$ when $W = .776$; for $W = .79$, P is just a little larger than .01 ($P = .05$ when $W = .842$, so perhaps $P = .012$).

13-R7 In view of the "research hypothesis," that high G tends to produce daughters, a 1-sided (2-sample) Z-test is appropriate:

$$Z = \frac{\frac{295}{582} - \frac{66}{166}}{\sqrt{\frac{361}{748}\cdot\frac{387}{748}\left(\frac{1}{582} + \frac{1}{166}\right)}} = 2.485, \quad P = .0065.$$

We could use the chi-square statistic [estimated expected cell frequencies are 280.9, 80.1, 301.1, 85.9], which turns out to be 6.17 (2.485^2), and the tail area beyond this (1 d.f.) is .0013. This is twice the P-value found using Z because the area under the single tail of chi-square is the area under the Z-density *both* to the left of -2.485 and to the right of 2.485. (That is, the chi-square test is a two-sided Z-test.)

13-R8 (You can leave off the integer 9 from each observation, since this does not affect the correlation to be calculated.) We used a computer to find $r^2 = .9284$, and Table XIV shows a P-value of about .10. To do the calculation with a hand calculator that does correlations, you first have to order the data, and then pair the ordered data with corresponding rankits.

13-R9 This is just a two by two contingency table, analyzed either with Z or with Pearson's chi-square. Since the alternative is apt to be one-sided we'd use Z (chi-square is for a two-sided alternative, as explained above in 13-R7). The estimate of the common value of the population proportions (under H_0) is the total number of successes divided by the total number of trials: 9/50:

$$Z = \frac{\frac{6}{23} - \frac{3}{27}}{\sqrt{\frac{9}{50} \cdot \frac{41}{50}\left(\frac{1}{23} + \frac{1}{27}\right)}} = 1.374, \quad P = .085.$$

CHAPTER 14

14-R1 Because the raw data are not available, we calculate sums of squares from the given statistics. First, the grand mean—a weighted average of the three sample means:

$$\overline{X} = \frac{35 \times 28 + 25 \times 35 + 22 \times 33}{35 + 25 + 22} = 31.476.$$

With this, we can find the treatment sum of squares:

$$\text{SSTr} = \sum n_i (\overline{X}_i - \overline{X})^2$$
$$= 35 \times 3.476^2 + 25 \times 3.524^2 + 22 \times 1.524^2 = 784.45.$$

The error mean square is the "pooled variance" estimate of σ^2:

$$\text{MSE} = \frac{34 \times 10^2 + 24 \times 10^2 + 21 \times 7^2}{34 + 24 + 21} = \frac{6829}{79} = 86.44.$$

The treatment d.f. is $k - 1 = 2$, and the error d.f. is $n - k = 82 - 3$:

Source	d.f.	SS	MS	F
Treatment	2	784.45	392.22	4.54
Error	79	6829	86.44	
Total	81	7613.45		

The last entry in Table VIIIa (numerator d.f. $= 2$) opposite 4.4 is for 40 d.f. in the denominator, from which we infer only that $P < .016$. (Our statistical software gives $P = .0135$.) There is some basis for believing that the population means are not all the same.

14-R2 The weights used in calculating $\overline{X}$ would have to be equal, which means that the sample sizes would be equal.

14-R3 The Scheffé method gives simultaneous intervals (for a given "error rate" α) for *all* contrasts, not just the possible pairwise comparisons. The intervals have to be a bit wider so that one claim that all contrasts will be covered by the calculated intervals with probability $1 - \alpha$.

14-R4 It can happen, even if there is no mistake in keying in the data, if you round off too severely. This can also happen in just calculating a sample variance using the formula $(n - 1)S^2 = \sum X_i^2 - n\overline{X}^2$. If you have a calculator that does s.d.'s, enter the data 9999990, 9999992, 9999994, and see what it gives for S. (It may give just 0, instead of reporting out a negative number.) The problem is that, working with a certain number of significant digits, the square of 9999990 would be rounded off. [If you use $\sum (X_i - \overline{X})^2$ for the r.h.s., there's no problem.]

14-R5 By definition, $\tau_i = \mu_i - \mu$, where $\mu = \frac{1}{n}\sum n_i\mu_i$. Then,

$$\sum_1^k n_i(\mu_i - \mu) = \sum_1^k n_i\mu_i - \mu\sum_1^k n_i = n\mu - n\mu.$$

14-R6 By definition, $\tau_i = \theta_i - \mu$, where $\mu = \frac{1}{rc}\sum_i\sum_j E(X_{ij})$, $\theta_i = \frac{1}{c}\sum_j E(X_{ij})$.

Then, $\sum_i(\theta_i - \mu) = \sum_i\frac{1}{c}\sum_j E(X_{ij}) - \sum_{i=1}^r \mu = r\mu - r\mu.$

(Showing $\sum_j \beta_j = 0$ is completely analogous.)

14-R7 (a) SSE $= 22S_p^2 = 15 \times 11.3^2 + 7 \times 8.3^2 = 2397.58$, and $\bar{X} = 34$, the

weighted average of sample means: $\dfrac{16 \times 40.3 + 8 \times 21.4}{24}$. And then,

SSTr $= \sum n_i(\bar{X}_i - \bar{X})^2 = 16(40.3 - 34)^2 + 8(21.4 - 34)^2 = 1905.12.$

The ANOVA table is:

Source	d.f.	SS	MS	F
Treatment	1	1905.12	1905.12	17.48
Error	22	2397.58	108.98	
Total	23	4302.70		

Doing it this way, we get a *P*-value (.0004) which corresponds to two tails of the corresponding *T* [in (b)].

(b) The pooled variance S_p^2 is the MS in (a), so the 2-sample *t*-statistic is

$$T = \frac{40.3 - 21.4}{\sqrt{108.98(1/16 + 1/8)}} = \frac{18.9}{4.52} = 4.18 = \sqrt{17.48}.$$

With 22 d.f., the *P*-value from Table IIIb is shown as .000. (Statistical software gives us the value .00019.)

14-R8 For any single cell, the sum of squared deviations about the cell mean divided by $m - 1$ is the cell variance; the sum of squared deviations divided by σ^2 is therefore chi$^2(m-1)$, because of the assumption of independent observations from a normal population. Adding all of these (independent) quantities (that is, over all cells), we have a sum of independent chi-square variables, which is chi-square—with degrees of freedom equal to the sum of the degrees of freedom of the summands: d.f. $= rc(m-1)$. Under the hypothesis that the cell means are equal, SSTotal is the sum of squared deviations of rcm observations about their mean, which, when divided by σ^2, is chi$^2(rcm - 1)$.

CHAPTER 15

15-R1 (b) $S_x^2 = .47574$, so $SS_{xx} = (n-1)S_x^2 = 12 \times .47574 = 5.7089$.

(c) Calculator yields: $\widehat{\beta} = -24.02$, $r^2 = .9697$, $S_Y^2 = 283.047$. Then,

$$SS_{YY} = 12S_Y^2 = 3396.56, \; SSE = SS_{YY}(1-r^2) = 102.85,$$

$$\widehat{\sigma}^2 = MSE = \frac{SSE}{11} = 9.350, \text{ and s.e.}(\widehat{\beta}) = \sqrt{\frac{\widehat{\sigma}^2}{SS_{xx}}} = 1.280.$$

90% confidence limits: $-24.02 \pm 1.8 \times 1.280$, or $-26.32 < \beta < -21.72$, where 1.8 is the 95th percentile of $t(11)$.

(d) The fraction explained as error is $1 - r^2 = 3.03\%$.

15-R2 There really isn't enough data to draw conclusions from; we include these data, in part because they were thought to be worth publishing, and in part to give the student an exercise in the methods without having to enter a lot of data.

(a) Let $x = $ log percent lithium, and $y = $ log of setting time:

x	y	$\widehat{y}$	$y - \widehat{y}$
-7.60	8.869	8.493	.376
-6.908	7.346	7.868	$-.522$
-5.298	6.522	6.416	.106
-4.605	5.829	5.790	.039

The equation of the least squares line is $y = 1.6358 - .9022\,x$ (done on a calculator), and for any x, the predicted value is $\widehat{y} = 1.6358 - .9022\,x$. We calculated this for each x in the data, entering it in the table as $\widehat{y}$.

(b) The residuals are shown in the table above. The plot has too few points to be able to infer much of anything with confidence. It is at least possible that the assumptions about the error components hold.

(c) The rankits (from Table XII) are -1.029, $-.297$, .297, 1.029, and the correlation with the ordered residuals is $r = .9542$, so $W = r^2 \doteq .91$. But don't put too much faith in this! With only 4 observations, any inference at all is risky, indeed.

15-R3 (a) The scatter plot is shown below.

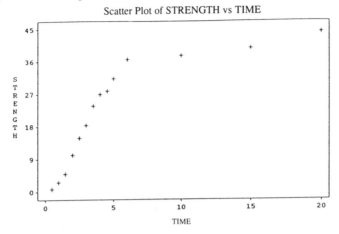

Scatter Plot of STRENGTH vs TIME

It seems clear that something happens around $x = 7$—what started out to be a nearly linear plot suddenly changes direction. It is pointless to fit a straight line to all the data.

(b) Without cases 12-14, the plot is shown in the following figure. The new correlation coefficient is $r = .992$. (This illustrates what influence points far away from the others can have on a correlation.)

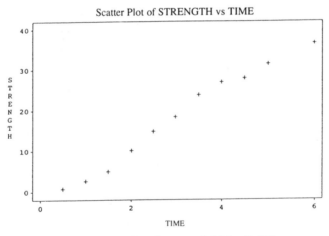

Scatter Plot of STRENGTH vs TIME

(c) The new least squares line is $y = -3.366 + 7.006\,x$.

(d) To predict Y, substitute $x = 5.5$ in the equation of (c): $y = 35.2$. Predicting Y well beyond the range of the data used in finding the line is always risky; here it is almost certain that the line that fits the first 11 cases is not going to give a good prediction where $x = 12$.

(e) The plot of standardized residuals is shown below. There seems to be an "up-down-up" tendency, but it is not clear enough to permit a definite conclusion.

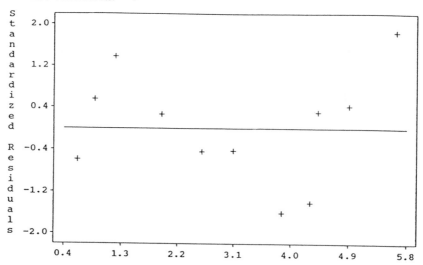

15-R4 (a) The scatter plot is as follows:

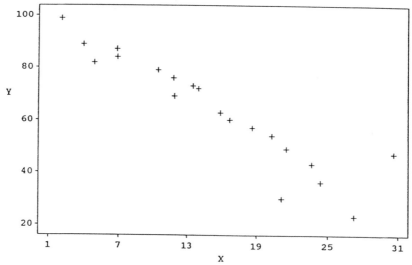

The least squares line is $y = 100.61 - 2.4156\, x$. We predict the number to be $100.61 - 2.4156 \times 15 = 64.4$. The standard error of prediction reported by the software we used is 8.16. We can calculate it from $r^2 = .8638$, $SS_{YY} = SSTotal = 8380.8$, $SSE = SS_{YY}(1 - r^2) = 1141.46$,

$\mathrm{MSE} = \mathrm{SSE}/\mathrm{d.f.} = 63.414$, and $\mathrm{SS}_{xx} = 19 \times S_x^2$:

$$\mathrm{m.s.p.e.} = \mathrm{MSE}\left\{1 + \tfrac{1}{n} + \frac{(x - \bar{x})^2}{\mathrm{SS}_{xx}}\right\} = 63.414\left\{1 + \tfrac{1}{20} + \frac{(15 - 15.32)^2}{19 \times 21^2}\right\}$$

$$= 66.585 = 8.16^2.$$

(b) The standardized residuals, plotted against x, are shown below:

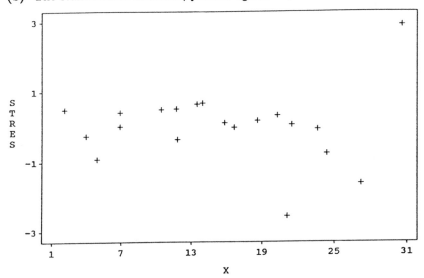

There seems to be an increase in variability with x, which is apparent also on the scatter plot given in (a) above. A nonconstant variance would invalidate the inferences in (a), if taken literally; but perhaps the conclusions there are close enough for practical purposes.

15-R5 $\sum[(Y_i - \bar{Y}) - \widehat{\beta}(x_i - \bar{x})]^2$

$$= \sum(Y_i - \bar{Y})^2 + \widehat{\beta}^2\sum(x_i - \bar{x})^2 - 2\widehat{\beta}\sum(x_i - \bar{x})(Y_i - \bar{Y})$$

$$= \mathrm{SS}_{YY} + \frac{\mathrm{SS}_{xY}^2}{\mathrm{SS}_{xx}^2}\cdot\mathrm{SS}_{xx} - 2\frac{\mathrm{SS}_{xY}}{\mathrm{SS}_{xx}}\cdot\mathrm{SS}_{xY}$$

$$= \mathrm{SS}_{YY}\left\{1 - \frac{\mathrm{SS}_{xY}^2}{\mathrm{SS}_{xx}\mathrm{SS}_{YY}}\right\} = \mathrm{SS}_{YY}(1 - r^2)$$

15-R6 The sum of the residuals is $\sum(Y_i - \alpha - \beta x_i)$. One of the two equations we derived as necessary to minimize the sum of squared residuals tells us that this sum is 0 when $\alpha = \widehat{\alpha}$ and $\beta = \widehat{\beta}$.

15-R7 The text notation was set up to allow for any number of predictors. Here we have only two, so we simplify the notation, calling the predictors x and z, and the response Y. The data then consist of n cases (x_i, z_i, Y_i). The sum of squared residuals of the responses Y about the plane $y = \alpha + \beta x + \gamma z$ is

$$R(\alpha, \beta, \gamma) = \sum (Y_i - \alpha - \beta x_i - \gamma z_i)^2.$$

The derivatives with respect to the parameters are

$$\frac{\partial T}{\partial \alpha} = -2 \sum (Y_i - \alpha - \beta x_i - \gamma z_i),$$

$$\frac{\partial T}{\partial \beta} = -2 \sum x_i (Y_i - \alpha - \beta x_i - \gamma z_i),$$

$$\frac{\partial T}{\partial \gamma} = -2 \sum z_i (Y_i - \alpha - \beta x_i - \gamma z_i),$$

We set these equal to 0, cancel the -2's, transpose the negative terms:

$$\sum Y_i = n\alpha + \beta \sum x_i + \gamma \sum z_i$$

$$\sum x_i Y_i = \alpha \sum x_i + \beta \sum x_i^2 + \beta \sum x_i z_i$$

$$\sum z_i Y_i = \alpha \sum z_i + \beta \sum x_i z_i + \beta \sum z_i^2$$

[Translation: $x_i = x_{1i}$, $z_i = x_{2i}$, $\alpha = \beta_0$, $\beta = \beta_1$, $\gamma = \beta_2$.]

15-R8 If the $\widehat{\beta}$'s in the solution are *linear* functions of the independent, normal responses, they are normally distributed—according to Property iv in §6.1. [Somewhat more is actually true (although we haven't gone into this): in the case of two β's, the joint distribution of the estimators $\widehat{\beta}$'s is bivariate normal; when there are more than two, the joint distribution is "multivariate normal"—a distribution we haven't studied.]

15-R9 The correlation is quite strong: $r \doteq .99$. The coefficient and ANOVA tables printed out as follows [using *Statistix*®: Analytical Software, PO Box 12185, Tallahassee, FL 32317]:

```
PREDICTOR
VARIABLES      COEFFICIENT     STD ERROR      STUDENT'S T        P         VIF
---------      -----------     ---------      -----------      ------      -----
CONSTANT         -0.07716       0.01378          -5.60         0.0003
T                 0.01176       3.951E-04        29.76         0.0000      16.6
TSQ              -2.486E-05     2.288E-06        -10.87        0.0000      16.6


R-SQUARED              0.9986    RESID. MEAN SQUARE (MSE)   2.390E-04
ADJUSTED R-SQUARED     0.9982    STANDARD DEVIATION            0.01546


SOURCE         DF        SS           MS          F          P
----------     ---    ---------    ---------    -----     ------
REGRESSION      2      1.49772      0.74886     3133.09    0.0000
RESIDUAL        9      0.00215      2.390E-04
TOTAL          11      1.49987
```

The scatter plot is shown with a fitted line and 95% confidence and prediction intervals:

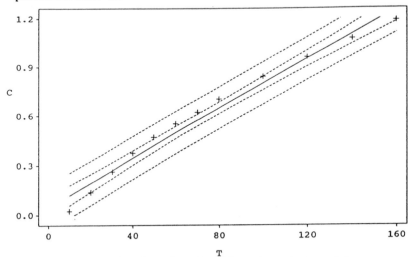

The fit is fairly tight—but there is a hint of regularity of the errors; this shows up more clearly in the residual plot:

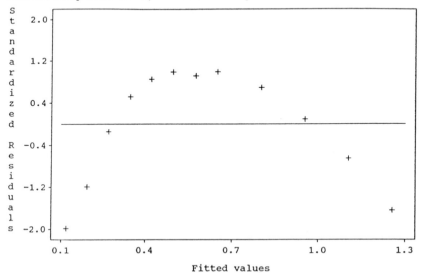

The residuals are not what we usually think of as i.i.d., let alone normal. We tried a quadratic fit, adding t^2 as a predictor:

PREDICTOR VARIABLES	COEFFICIENT	STD ERROR	STUDENT'S T	P
CONSTANT	0.04164	0.02991	1.39	0.1940
T	0.00760	3.453E-04	22.00	0.0000

R-SQUARED	0.9798	RESID. MEAN SQUARE (MSE)	0.00304
ADJUSTED R-SQUARED	0.9777	STANDARD DEVIATION	0.05511

SOURCE	DF	SS	MS	F	P
REGRESSION	1	1.46950	1.46950	483.85	0.0000
RESIDUAL	10	0.03037	0.00304		
TOTAL	11	1.49987			

The correlation is nearly 1. Surprisingly, however, the residual plot again suggest some regularity:

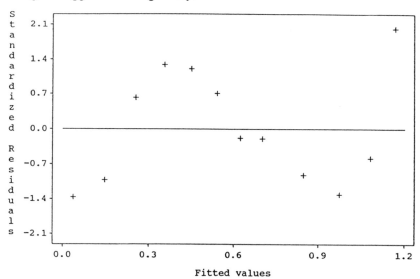

Fitted values

Curious, we added a cubic predictor. The fit is incredibly close:

PREDICTOR VARIABLES	COEFFICIENT	STD ERROR	STUDENT'S T	P	VIF
CONSTANT	-0.11951	0.00766	-15.60	0.0000	
T	0.01447	3.908E-04	37.01	0.0000	115.8
TSQ	-6.513E-05	5.448E-06	-11.95	0.0000	671.0
TCU	1.611E-07	2.152E-08	7.48	0.0001	260.6

R-SQUARED	0.9998	RESID. MEAN SQUARE (MSE)	3.360E-05
ADJUSTED R-SQUARED	0.9998	STANDARD DEVIATION	0.00580

SOURCE	DF	SS	MS	F	P
REGRESSION	3	1.49960	0.49987	14876.90	0.0000
RESIDUAL	8	2.688E-04	3.360E-05		
TOTAL	11	1.49987			

The residual plot now shows no regularity—it looks more like what a plot of residuals ought to look like when the assumptions are satisfied.

What we've been doing should not really be carried out in isolation, apart from the investigators whose insight and knowledge of the subject matter would better guide the analysis to a useful conclusion.

15-R10   The correlation coefficient (hand calculator result) is $r = .1901$, with $r^2 = .03615$. With these values in the t-statistic for testing $\rho = 0$, we obtain

$$T = \sqrt{n-2}\,\frac{r}{\sqrt{1-r^2}} = \sqrt{8}\,\frac{.190}{\sqrt{.96385}} \doteq .55.$$

Referring to Table IIIa (percentiles of the t-distribution), we see that the area beyond .55 (8 d.f.) is $P \doteq .30$. A sample correlation of .19 is not far enough from 0 (with a sample of this size) to cast doubt on $\rho = 0$. This is not to say that ρ actually is 0; simply, that we have no evidence of that hypothesis in the present sample.